AF598103

ADVANCES IN EBOLA CONTROL

Edited by **Samuel Ikwaras Okware**

Advances in Ebola Control
http://dx.doi.org/10.5772/intechopen.69364
Edited by Samuel Ikwaras Okware

Contributors

Sulaiman Bah, Ehsan Raoufi, Maryam Hemmati, Hossein Fallahi, Rachele Hwong, Wen-Ta Chiu, Jonathan Wu, Stanley Toy, Jr., Jj Stewart, Jennifer Chang, Usman Sumo Friend Tambunan, Mochammad Arfin Fardiansyah Nasution, Ahmad Husein Alkaff, Kasangye Kangoy Aurelie, Mutangala Muloye Guy, Ngoyi Fuamba Bona, Kaya Mulumbati Charles, Patrick Mawupemor Avevor, Li Shixue

Notice

First published in London, United Kingdom, 2018 by IntechOpen
IntechOpen is the global imprint of INTECHOPEN LIMITED, registered in England and Wales, registration number: 11086078, The Shard, 25th floor, 32 London Bridge Street
London, SE19SG – United Kingdom
Printed in Croatia

British Library Cataloguing-in-Publication Data
A catalogue record for this book is available from the British Library

Additional hard copies can be obtained from orders@intechopen.com

Advances in Ebola Control, Edited by Samuel Ikwaras Okware
p. cm.
Print ISBN 978-1-78923-070-3
Online ISBN 978-1-78923-071-0

For EU product safety concerns:
IN TECH d.o.o., Prolaz Marije Krucifikse Kozulić 3, 51000 Rijeka, Croatia,
info@intechopen.com or visit our website at intechopen.com.

Meet the editor

Dr. Samuel Ikwaras Okware is a medical doctor and public health specialist. He obtained his PhD degree in Ebola and Emerging Infections from the University of Bergen. He is currently the director general of the Uganda National Health Research Organisation, which coordinates national health research. Dr. Okware has many years of experience in public health and communicable disease control. He led the containment of five major Ebola outbreaks in Uganda between 2000 and 2012. Earlier, he also contributed significantly to the pioneering work and research on HIV/AIDS, which led to early declines in HIV prevalence in the country. He has international experience as a public health consultant and was a member of several advisory committees including the WHO Expert Committee on Research and Development.

Contents

Preface

Ebola virus disease is a serious transmissible infection, with a high propensity for outbreaks, associated with bleeding tendencies, multiple organ failure, and high fatality within days. It has no cure. The devastating outbreak in 2014 in West Africa reported 28,000 cases with 11,000 deaths, most of whom were healthcare workers. Health service delivery was greatly undermined. The subsequent disease outcomes had profound societal and economic impact on communities and generated unprecedented emergency and panic across the world.

This book examines some current emerging advances on Ebola and its management. Contributions to this book have been made by scientists from various parts of the world with knowledge and expertise in various specific fields of Ebola control.

The first section discusses selected aspects of Ebola surveillance. It provides the historical perspective, which gives a platform to discuss the origins and geographical scope and details of taxonomy and evolution of Ebola virus disease. It examines the global experiences influencing preparedness, prevention, detection, and response. It also discusses emerging reservoirs, diagnostics, and plausible modes of spread in at-risk populations. The need for stronger collaboration and partnerships is emphasized to accelerate the capacity for early detection and surveillance. New approaches to improve surveillance are examined with the possibility of registering the disease as a multiple cause of death (part of the civil registration/vital statistic system), against a framework of a scoring system to link the two. The chapter discusses the steps that African countries could take to achieve a functional system that records multiple causes of death.

The second section focuses on the management of Ebola in high-risk situations. It explores the challenges of a pregnant Ebola survivor in delivery of a baby. A case is made in which a safe delivery was possible in a county hospital. Recommendations and standard operating procedures for safe delivery are described.

The third section focuses on the current advances in the development of drugs and vaccines for Ebola. One chapter reviews the peptide epitopes for superficial glycoprotein (GP) and nucleoprotein (NP) of *Zaire Ebola virus* with the use of in silico methods and immunoinformatic approach. The resulting peptides could then stimulate immune response and may contribute to designing multipeptide vaccines and drugs for prevention and screening of Ebola.

I thank the authors for their contributions. I thank Edi Lipović, the project manager, for his invaluable support and assistance. I am also grateful for the support extended to the editor by my team at the Uganda National Health Research Organisation.

Dr. Samuel Ikwaras Okware, PhD
Director General
Uganda National Health Research Organisation
Entebbe, Uganda

Section 1

Surveillance

Chapter 1

A Historical Review of Ebola Outbreaks

Kasangye Kangoy Aurelie, Mutangala Muloye Guy, Ngoyi Fuamba Bona, Kaya Mulumbati Charles, Avevor Patrick Mawupemor and Li Shixue

Additional information is available at the end of the chapter

http://dx.doi.org/10.5772/intechopen.72660

Abstract

Ebola Virus Disease (EVD) is a severe, often fatal illness in humans caused by the Ebola virus. Since the first case was identified in 1976, there have been 36 documented outbreaks with the worst and most publicized recorded in 2014 which ravaged three West African Countries, Guinea, Liberia and Serial Leone. The West African outbreak recorded 28,616 human cases, 11,310 deaths (CFR: 57–59%) and left about 17,000 survivors, many of whom have to grapple with Post-Ebola syndrome. Historically, ZEBOV has the highest virulence. Providing a historical perspective which highlights key challenges and progress made toward detecting and responding to EVD is a key to charting a path towards stronger resilience against the disease. There have been remarkable shifts in diagnostics, at risk populations, impact on health systems and response approaches. The health sector continues to gain global experiences about EVD which has shaped preparedness, prevention, detection, diagnostics, response, and recovery strategies. This has brought about the need for stronger collaboration between international organizations and seemingly Ebola endemic countries in the areas of improving disease surveillance, strengthening health systems, development and establishment of early warning systems, improving the capacity of local laboratories and trainings for health workers.

Keywords: Ebola, outbreaks, world

1. Introduction

Ebola Virus Disease (EVD), formerly known as Ebola Hemorrhagic Fever (EHF) is a severe, often fatal illness in humans [1]. It has become well known and notified disease all over the world, since its last outbreak in Guinea, Sierra Leone and Liberia (December 2013). EVD is caused by the Ebola virus and is responsible for about 50–90% death in clinically diagnosed

cases [2]. Efforts to contain this disease have been the focus of the World Health Organization (WHO) and some other countries in recent times. Despite these efforts, no medicine has yet been licensed for the treatment of the disease [3]. The Ebola virus was first discovered in Zaire now called the Democratic Republic of Congo (DRC). The virus was named Ebola following the first outbreak in the town of Yambuku, which is near the Ebola River in DRC; it is at the hospital in this town that the first case of Ebola was identified in September 1976 by the Belgian doctor Peter Piot of the Institute of Tropical Medicine Anvers [4, 5]. This study aims to summarize old and new experiences of Ebola all over the world, in order to have an overview of all Ebola outbreaks and to propose strategies for better prevention and management of future outbreaks.

1.1. Etiology

The Ebola virus (EBOV) is the principal etiology of EVD [6], Ebola virus belongs to the family of Filoviridae, to the order of Mononegavirales which includes Rhabdoviridae and Paramyxoviridae. The virion is pleomorphic, producing "U"-shaped, "6"-shaped, or circular forms but the predominant forms of the virion most frequently seen by electron microscope are long tubular structures. It contains one molecule of linear, single-stranded, negative-sense RNA of 4.2×10^6 Da [7]. EVD is caused by five genetically distinct members of the Filoviridae family:

1. Zaire ebolavirus (ZEBOV): Up to 2000, Ebola virus (EBOV) was formerly designated by Zaire Ebola virus [8, 9]. And in 2002, to species Zaire ebolavirus [10, 11]. However, most scientific articles continued to refer to "Ebola virus" or used the terms Ebola virus and Zaire ebolavirus in parallel. Consequently, in 2010, a group of researchers recommended that the name "Ebola virus" be adopted for a subclassification within the species Zaire ebolavirus, with the corresponding abbreviation EBOV [12]. Previous abbreviations for the virus were EBOV-Z (for Ebola virus Zaire) and ZEBOV (for Zaire Ebola virus or Zaire ebolavirus). At present, ICTV does not officially recognize "Ebola virus" as a taxonomic rank, and rather continues to use and recommend only the species designation Zaire ebolavirus [13].

2. Sudan ebolavirus (SEBOV): Sudan virus was first introduced as a new "strain" of Ebola virus in 1977 [14]. Sudan virus was described as "Ebola haemorrhagic fever" in a 1978 WHO report describing the 1976 Sudan Ebola outbreak [15]. In 2000, it received the designation of Sudan Ebola virus [8, 9], and in 2002 the name was changed to Sudan ebolavirus [10, 11]. Previous abbreviations for the virus were EBOV-S (for Ebola virus Sudan) and most recently SEBOV (for Sudan Ebola virus or Sudan ebolavirus). The virus received its final designation in 2010, when it was renamed Sudan virus (SUDV) [12].

3. Côte d'Ivoire ebolavirus (CEBOV): The name Taï Forest ebolavirus is derived from Parc National de Taï (the name of a national park in Côte d'Ivoire, where Taï Forest virus was first discovered) and the taxonomic suffix ebolavirus (which denotes an ebolavirus species) [12]. Taï Forest virus was first introduced as a new "strain" of Ebola virus in 1995 [16]. In 2000, it received the designation Côte d'Ivoire Ebola virus [8, 9]. In 2002, the name was changed to Cote d'Ivoire ebolavirus [10, 11]. The virus received its final designation in 2010, when it was renamed Taï Forest virus (TAFV) [12].

4. Bundibugyo ebolavirus (BEBOV): Bundibugyo virus was first introduced as Bundibugyo ebolavirus in 2008 [16]. The name Bundibugyo virus is derived from Bundibugyo (the name of the chief town of the Ugandan Bundibugyo District, where it was first discovered) and the taxonomic suffix virus [12]. Another name introduced at the same time was Uganda ebolavirus [17]. Later publications also referred to the virus as a novel "strain" of Ebola virus [18], or as Bundibugyo Ebola virus [19]. The abbreviations BEBOV (for Bundibugyo ebolavirus) and UEBOV (for Uganda ebolavirus) [17], were briefly used before BDBV was established as the abbreviation for Bundibugyo virus [12].

5. Reston ebolavirus (REBOV): Reston virus was first introduced as a new "strain" of Ebola virus in 1990 [20]. In 2000, it received the designation Reston Ebola virus [26, 27], and in 2002 the name was changed to Reston ebolavirus [8, 9]. Previous abbreviations for the virus were EBOV-R (for Ebola virus Reston) and most recently REBOV (for Reston Ebola virus or Reston ebolavirus). The virus received its current designation in 2010, when it was renamed Reston virus (RESTV) [12].

1.2. Transmission

Transmission in most outbreaks, Ebola virus is introduced into human populations via the handling of infected animal carcasses. In these cases, the first source of transmission is an animal found dead or hunted in the forest, followed by person-to person transmission from index case to family members or health-care staff. Animal-to-human transmission occurs when people come into contact with tissues and bodily fluids of infected animals, especially with infected NHPs [21]. The most likely vector of the EBOV is the fruit bat, specifically *Hypsignathus monstrosus* (the hammer-headed fruit bat), *Epomops franqueti* (Franquet's epaulets fruit bat), and *Myonycteris torquata* (the little-collared bat) [22].

Between 1976 and 1998, in 30,000 mammals, birds, reptiles, amphibians and arthropods sampled from regions of EBOV outbreaks, no Ebola virus was detected apart from some genetic traces found in six rodents (belonging to the species *Mus setulosus* and *Praomys*) and one shrew (*Sylvisorex ollula*) collected from the Central African Republic [23, 24]. Further research efforts have not confirmed rodents as a reservoir [25]. Traces of EBOV were detected in the carcasses of gorillas and chimpanzees during outbreaks in 2001 and 2003, which later became the source of human infections. The high rates of death in these species resulting from EBOV infection make it unlikely that these species represent a natural reservoir for the virus [23]. Antibodies against Zaire and Reston viruses have been found in fruit bats in Bangladesh, suggesting that these bats are also potential hosts of the virus and that the filoviruses are present in Asia [26].

The means of transmission within bat populations remain unknown [27]. Human disease is thought to result from consumption of poorly-cooked infected animals, such as bats or chimpanzees (which are known to feed on bats) [22, 28]. According to the findings of the WHO in October 2014, the most infectious fluids are blood, feces and vomit. The virus has also been detected in breast milk and urine [29]. However unlike other zoonosis, Ebola has the potential of spreading from human to human through exposure of mucous membranes or broken skin to infected body fluids including large aerosol droplets that can be produced during coughing [30].

1.3. The clinical features

The clinical features can be divided into four main phases as follows:

(Phase A) Influenza–like syndrome: The onset is abrupt with non-specific symptoms or signs such as high fever, headache, arthralgia, myalgia, sore throat, and malaise with nausea.

(Phase B) Acute (day 1–6): Persistent fever not responding to antimalarial drugs or to antibiotics, headache, and intense fatigue, followed by diarrhea and abdominal pain, anorexia and vomiting.

(Phase C) Pseudo-remission (day 7–8): During this phase the patient feels better and seeks food. The health situation presents with some improvement. Some patients may recover during this phase and survive from the disease.

(Phase D) Aggravation (day 9): respiratory disorders (dyspnea, throat and chest pain, cough, hiccups), symptoms of hemorrhagic diathesis (bloody diarrhea, hematemesis, conjunctival injection, gingival bleeding, nosebleeds and bleeding at the site of injection consistent with disseminated intravascular coagulation), skin manifestations (petechiae, purpura, morbilliform skin rash), neuropsychiatric manifestations (prostration, delirium, confusion, coma) and cardio-vascular distress and hypovolemic shock (death) [7].

Patients do not transmit Ebola during the incubation period but become infectious once they develop clinical features of EVD [30]. From the clinical manifestations it is obvious that EVD may mimic many other tropical diseases like malaria, typhoid fever or yellow fever at the start of the disease. In most outbreaks, recognition of the disease is delayed because physicians are not accustomed to this new illness and the symptoms are generally non-specific. Outside the epidemic context, it appears quite impossible to recognize the first Ebola case in an outbreak on clinical grounds only. Suspicion of EVD is only possible later during the aggravation phase [7].

2. Methodology

This study aims to summarize results of publications on all Ebola outbreaks. In order to accomplish this work, information was taken from databases such as PubMed and Cochrane library, and some articles were also taken from Google Scholar. This search will focus on past and present Ebola outbreaks all over the world. For some abstracts that met the predefined inclusion criteria, full texts were obtained. The data collection was focused more on some aspects of each outbreak such as: the year of the outbreak, the geographical spread (estimated area covered by the outbreak, country and region), and the strain of the virus involved in each outbreak, the index case, the case fatality, the diagnosis and the treatment used to control the situation. All the data will be put in Microsoft Excel software for construction of graphs. The data collection started on first July and ended on first August 2017; and a total of 23 full text and 6 abstracts were selected for the data extraction.

http://dx.doi.org/10.5772/intechopen.72660

3. Ebola outbreaks characteristics (1976–2017)

This section will focuses on the characteristics of all Ebola outbreaks. In total, there have been 36 documented Ebola outbreaks that can be grouped into two: Major/Massive cases and Minor/Single cases.

3.1. Major outbreaks

Major outbreaks are larger outbreaks with more than 10 human registered cases of EVD (19 outbreaks).

The table highlights each major outbreak, the viral species responsible for the outbreak with the specie that induce the EVD, the country and the year in which the outbreak occurred, and the number of cases and deaths recorded. **Table 1** shows also after the West African Ebola Epidemic, Uganda is the second country in terms of number of cases, it has registered a lot of cases of Ebola during its first outbreak of 2000–2001 (425 cases/224 deaths).

3.2. Minor outbreaks

Minor/Single cases: these are smaller outbreaks with less than 10 human cases of EVD (17 outbreaks in total) (**Table 2**: minor Ebola outbreaks).

The DRC (Zaire) has recorded the highest number of EVD outbreaks (8 in total). It is also important to note that some of the Ebola cases were asymptomatic in minor outbreak such as in the Philippines 1, Philippines 3 and USA 2 outbreaks.

3.3. Case fatality rate of Ebolavirus

A case fatality rate (CFR) or case fatality risk is a property of an infectious disease in a particular population which states the risk of fatality due to the disease per case [31].

Figure 1 shows the distribution of case fatality by outbreaks (major outbreaks).

Figure 1 shows that the highest case fatality of major EVD outbreak in the all story of Ebola was recorded in the first outbreak in the Republic of Congo caused by ZEBOV (90%). and the lowest case fatality occurred in the fourth Ugandan outbreak caused by SUDV (34%). This corroborated with Literature that has reported Zaire species to have a higher case fatality than Sudan and Bundibugyo species, case fatality rates for ZEBOV as high as 90% [32].

3.4. Distribution of outbreaks by species of Ebola virus

Figure 2 shows how often each species of Ebola virus has been observed in the registered outbreaks. **Figure 2** also shows that ZEBOV is most commonly reported specie responsible for Ebola outbreaks.

Name*	Year	Cases/deaths	Country (city)/strain
Zaire1	August 1976	318/280	Zaire (Democratic Republic of Congo/DRC) in Yambuku/ZEBOV
Sudan1	November 1976	284/151	Sudan occurred in Nzara, Maridi, Tumbura and Juba/SUDV
Sudan2	1979	34/22	Sudan occurred in Nzara and Maridi/SUDV
Gabon1	1994	52/31	Gabon occurred in Makokou/ZEBOV
Zaire3	1995	315/254	Zaire in Kikwit/ZEBOV
Gabon2	1996 (January to April)	37/21	Gabon in Mayibout area/ZEBOV
Gabon3	1996–1997 (July to January)	60/45	Gabon occurred in Booue area/ZEBOV
Uganda1	2000–2001	425/224	Uganda in the Gulu, Masindi, and Mbarara district/SUDV
Gabon4	2001–2002 (October to July)	135/107	Gabon and Republic of the Congo/ZEBOV
Congo1	2002–2003(December to April)	143/128	Republic of Congo in the district of Mbomo and Kelle/ZEBOV
Congo2	2003 (November to December)	35/29	Republic of Congo occurred in Mbomo and Mbandza/ZEBOV
Sudan3	2004	17/7	Sudan in Yambio/SUDV
DRC1	2007	264/187	DRC in Kasai-Occidental province/ZEBOV
Uganda2	2007–2008 (December to January)	149/47	Uganda in the Bundibugyo district/BDBV
DRC2	2008–2009 (December to Febuary)	32/14	DRC occurred in Mweka and Luebo/ZEBOV
Uganda4	2012 (June to August)	24/17	Uganda in Kibaale district/SUDV
DRC3	2012 (June to November)	77/36	DRC in the Orientale Province/BDBV
West Africa	2013–2016	28.161/11,310	West African Ebola Virus Epidemic: Liberia, Sierra Leone, Guinea It began in Gueckedou (Guinea) in December 2013 ZEBOV
DRC4	2014 (August to October)	66/49	DRC in the Equateur Province/ZEBOV

Notes: Sudan here refers to South Sudan, formerly Sudan.

*Chronological Name of outbreak with the country name as the prefix and the number of time that outbreak occurred in that country as the suffix.

Table 1. Major Ebola outbreaks.

3.5. The West African outbreak (December 2013–2016)

The West African EVD outbreak still and remains the most severe and largest outbreak. It has divested 3 principal countries: Liberia, Sierra Leone and Guinea, and spread abroad. Small outbreaks occurred in Nigeria and Mali [33, 34], and isolated cases were recorded in Senegal [35],

http://dx.doi.org/10.5772/intechopen.72660

Name*	Year	Number of reported cases/number of death	Country (city)/strain
UK1	1976	1 case/0 death	United Kingdom/ZEBOV or SUDV
Zaire2	1977	1 case/1 death	Zaire in Tandala/ZEBOV
Philippines1	1989–1990	3 cases (asymptomatic)/0 death	Philippines/RESTV
USA1	1989	0 case/0 death	United States/RESTV
USA2	1990	4 cases (asymptomatic)/0 death	United States/RESTV
Italy1	1992	0 case/0 death	Italy/RESTV
C.I 1	1994	1 case/0 death	Cote d' Ivoire in Tai National Park/TAFV
C.I 2	1995	1 case/0 death	Cote d' Ivoire
SAF1	1996	2 cases/1 death	South Africa/ZEBOV
USA3	1996	0 case/0 death	United States/RESTV
Philippines2	1996	0 case/0 death	Philippines/RESTV
Russia1	1996	1 case/1 death	Russia/ZEBOV
Phillippines3	2008	6 cases (asymptomatic)/0 death	Philippines/RESTV
Uganda3	2011	1 case/1 death	Uganda in Luwero district/SUDV
Uganda5	2012–2013	6 cases/3 death	Uganda in Luwero district/SUDV
Phillipines4	2015	0 case/0 death	Philippines/RESTV
DRC5	2017	8 cases/4 death	DRC/ZEBOV

*Chronological Name of outbreak with the country name as the prefix and the number of time that outbreak occurred in that country as the suffix (e.g., Uganda 3 means the third outbreak in Uganda).

Notes: RESTV usually cause EVD in primates and others animals such as pigs, that why some of the outbreaks with RESTV as specie had 0 human cases and 0 deaths means that it was found in animals and not in human. However, Human being can get EVD with RESTV and not developed symptoms (that is the case of asymptomatic human cases of EVD).

Table 2. Minor Ebola outbreaks.

the United Kingdom and Sardinia [36, 37]. In addition, imported cases led to secondary infection of medical workers in the United States and Spain but did not spread further [38, 39].

Figure 3 shows the location of the West African Ebola outbreak.

It began in Guéckédou (Guinea) in December 2013 [41], On 25 March 2014 the WHO indicated that Guinea's Ministry of Health had reported an outbreak of Ebola virus disease in four southeastern districts, and that suspected cases in the neighboring countries of Liberia and Sierra Leone were being investigated [42], and on 29 March 2016, the WHO terminated the Public Health Emergency of International Concern status of the outbreak [43–45]. 28,616 human reported cases and 11,310 human deaths were registered with a case fatality of 57–59% (Among hospitalized patients [46, 47]. The outbreak left about 17,000 survivors of the disease, many of whom report post-recovery symptoms termed post-Ebola syndrome, often severe enough to require medical care for months or even years [48].

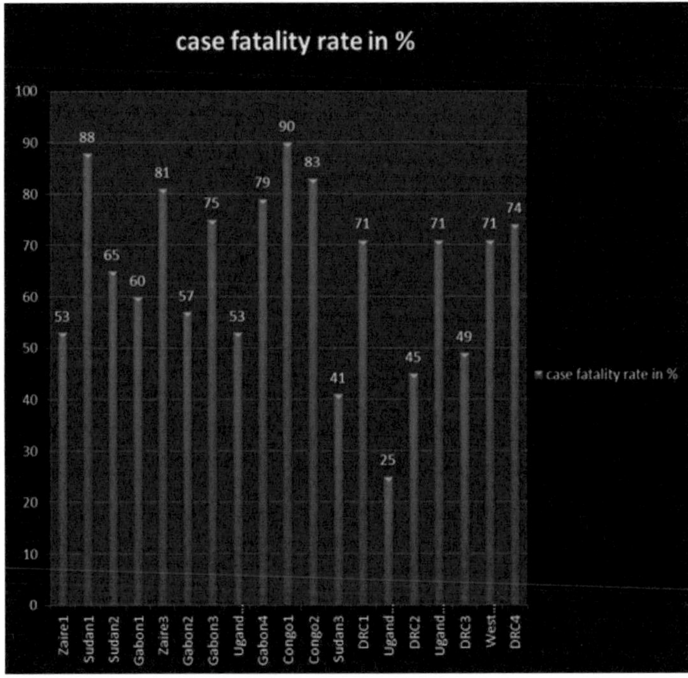

Figure 1. Distribution of case fatality rate by major outbreaks.

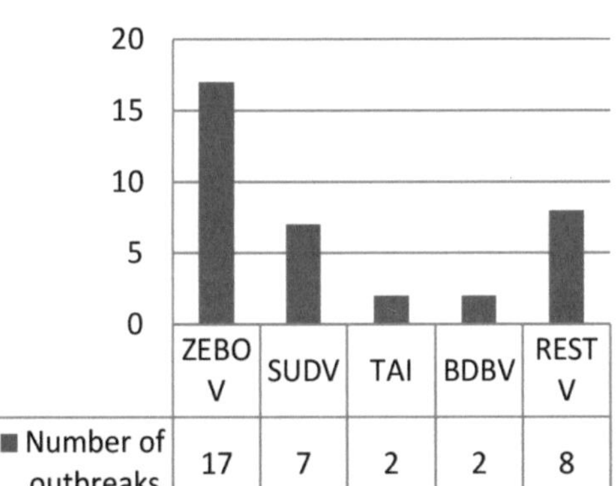

Figure 2. Distribution of outbreaks by species of Ebola virus.

Table 3 shows that Liberia registered a high number of Ebola cases as the number of deaths in the West African outbreak, however the high case fatality (in major outbreaks) was reported in Guinea.

It is worth noting that Nigeria was the first West African country to be declared Ebola free (20 October 2014) [49].

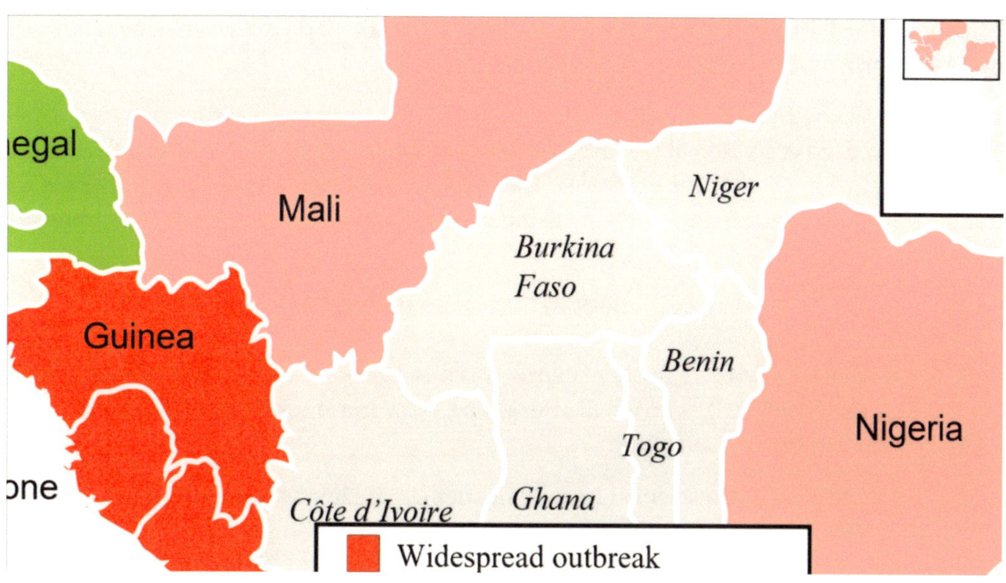

Figure 3. West African Virus Epidemic map [40].

Country	Number of cases	Number of deaths	Case fatality
Liberia*	10,666	4806	45%
Sierra Leone*	14,122	3955	28%
Guinea*	3804	2536	66%
Nigeria	20	8	40%
Mali	8	6	75%
USA	4	1	25%
Italy	1	0	0
UK	1	0	0
Senegal	1	0	0
Spain	1	0	0
Total	28,616	11,310	

*Major West African outbreak (Guinea, Liberia and Sierra Leone).

Table 3. Distribution of reported cases by countries in West African Ebola Epidemic.

3.6. Index cases

The index case, primary case, or patient zero is the initial patient in the population of an epidemiological investigation [50, 51]. The index case may indicate the source of the disease, the possible spread, and which reservoir holds the disease in between outbreaks. The index case is the first patient that indicates the existence of an outbreak. Earlier cases may be found and are labeled primary, secondary, tertiary, etc. [52].

In most of EVD outbreaks the index cases have to be in contact of a virus reservoir, eat an animal found dead or hunted in the forest, or also a traveler who was in contact with an Ebola case (Medical professionals for example). The index case of EVD is the point on which the human to human transmission starts; he is the bridge between animal and human transmission of the disease.

A Chronological list of some index cases in the history of Ebola:

1. The first recorded victim of the Ebola virus was a 44-year-old schoolteacher named Mabalo Lokela (in Zaire/DRC), who felt ill after eating fresh and smoked antelope and monkey; he died on 8 September 1976 [53].
2. In 1994 (Cote d' Ivoire) a scientist became ill after conducting an autopsy on a wild chimpanzee in the Tai Forest. The patient was treated in Switzerland [16].
3. In 1995 (Zaire) the index case was farming and preparing charcoal in the remnant forest areas of Kikwit, there were a lot of bats and rodents in the region [54].
4. In 1996 (January-April) in Mayibout area (Gabon), a chimpanzee found dead in the forest was eaten by people hunting for food. Nineteen people who were involved in the butchery of the animal became ill [55].
5. In 1996 (South Africa) a medical professional traveled from Gabon to Johannesburg, after having treated Ebola-infected patients and having been exposed to the virus. He was hospitalized, and a nurse who took care of him became infected and died [56].
6. 1996–1997 (Gabon) Occurred in Booué area with transport of patients to Libreville. Index case-patient was a hunter who lived in a forest camp. Disease was spread by close contact with infected persons. A dead chimpanzee found in the forest at the time was determined to be infected [55].
7. In 2000–2001 (Uganda): a farmer in Rwot Obillo village, 14 kilometers North of Gulu town was the index case [57].
8. In 2007 (DRC/Zaire) In Mweka, Kasai Occidental Province. The index case was the chief of the village and a hunter [58].
9. In 2011 (Uganda): On the 5th of May, a 13-year-old girl was admitted to Bombo hospital, the Sudan Ebola subtype was detected and confirmed [59].
10. In 2012 (Uganda): The index case was a 16-year-old female from a remote rural community. She fell sick while preparing forest land with her husband for the planting season. Nine relatives who participated at the funeral died including a mother, and several sisters who contracted the infection died [59].
11. In 2013 (West African outbreak), 2-year-old Emile Ouamouno is believed to be the index patient in the 2014 Ebola epidemic in Guinea and West Africa [60]. Scientists have long believed that bats are involved in the spread of the virus, and, incidentally, the boy's home

was in the vicinity of a large colony of Angolan free-tailed bats. The Ebola virus was, however, not found in any of the bats that were captured and tested [61]. His mother, sister, and grandmother later became ill with similar symptoms and also died; people infected by these initial cases spread the disease to other villages [62, 63].

12. In 2014 (DRC/Zaire) in the Equator province, the index patient was a pregnant woman living in Inkanamongo village, who butchered a monkey [64].

13. 2017 (DRC/Zaire), the first patient to be seen was a 39-year-old man who reported to the local health facility on 22nd April 2017. He was immediately referred to Likati health zone facility but he died in transit. On 24th April 2017, a motorcycle rider (who transported the first patient) and another person who supported the first patient during transportation developed acute febrile illness. The motor cycle rider subsequently died on 26 April 2017. Other people who were close to these patients eventually developed similar illness [65].

3.7. Ebola virus in health care workers

Morbidity and mortality caused by EVD among health care workers has been very important. The major difference between the management of the Ebola epidemic and others, such as the HIV epidemic, is that the Ebola virus presents a more challenging health hazard to health care providers. Nurses, doctors, Red Cross volunteers, and other health care workers stand the risk of being infected with the Ebola virus while providing care. The risk of EVD contamination among these health care workers is also increased in a continent like Africa where the nurses and other health care providers work under extraordinarily difficult conditions, lacking such basic infection control tools as bleach, soap, and gloves [2]. When an Ebola patient, comes with non-diagnosed EVD in a hospital, the chain of contamination can start with the health care provider that offers the first care [3].

The first and famous example of a contaminated health worker is the nurse Mayinga N'Seka, who died in the 1976 outbreak in Zaire (now DRC) and to whom the prototype Ebola virus variant Mayinga (EBOV/May) was named [66]. The 1995 Democratic Republic of the Congo (DRC) outbreak devastated health care workers, out of the 250 individuals who died, 47 (approximately 20%) were health care professionals [58]. In Uganda, in the first outbreak of 2000–2001, the were 31 health workers among victims; And in the 2007 outbreak, 14 health care workers were among the victims [59]. Another example is the case of a Congolese (DRC) doctor and three health workers, who undertook a postmortem cesarean section on the index case of the 2014 outbreak. Both were not only infected and died; but became the evident source of further transmission in this outbreak. And from that outbreak, there were 49 registered deaths, of which 8 were health care professionals [64].

In the West African outbreak, it was estimated that, depending on their occupation in the health service, health workers were between 21 and 32 times more likely to be infected with Ebola than people in the general adult population. WHO estimated that large number of nurses and nurse aides have been affected, accounting for more than 50% of all health worker infections with occupation reported. Other categories of health workers affected include medical workers (doctors and medical students (12%), laboratory workers and trade and elementary workers (janitors, maintenance staff, etc.) with 7% each [67]. In a study done only in

Guinea in 2015, among Guinean health care workers, incidence of Ebola infection was highest among laboratory technicians (34.7 per 1000) and doctors (26.6 per 1000), followed by midwives (8.7 per 1000) and nurses (5.5 per 1000) [68].

Many other health care workers have been contaminated while taking care of EVD patients in Africa and imported to other continents, thereby becoming the index cases for those countries. Examples are the United Kingdom (Glasgow, 2014), where a nurse coming from Sierra Leonne was considering a first case of Ebola to be diagnosed on British soil [69] and in USA (Texas, 2014), a healthcare worker coming from Liberia, reportedly a female nurse at Texas Presbyterian Hospital was the first known person-to-person transmission case of Ebola in the US [70].

In Africa, especially in rural area, Ebola outbreaks have been linked to many rumors and legends. The existence of rumors and legends related to the outbreaks could obscure the viral nature of the disease [7], and this can lead to difficulty, for health workers, to easily accomplish their tasks. Nurses and doctors had to deal with not only a panicked and fearful public, essentially absent public health and medical resources, but also they themselves were seen as agents of death [2]. In Kikwit (DRC), anyone associated with Ebola was likely to have experienced stigmatization. At a point during the outbreak, local people thought Ebola originated from the medical staff working in the hospital. All those who had died had been in a hospital. Therefore, the people reasoned that, it was the health care workers who were killing the people [2]. In West African countries some patients were taken to traditional healers rather than science in a bid to combat the disease [71, 72]; increasing then the risk of contamination in the population.

3.7.1. *Example of the experience of a health worker in Ebola outbreaks*

Doctor Bona Ngoyi one of the co-authors, who has provided health care to EVD patientS in three outbreaks: firstly in the 2014 outbreak in DRC, secondly in the West African outbreak in Guinea (2015) and thirdly in the last outbreak (2017) in the DRC- reported that: "*the general objective of the mission was to provide technical support in the fight against Ebola in all outbreaks. But each outbreak faces different challenges. For example, when i was assigned to the prefecture of Dubréka in Guinea (11 March 2015 to 10 May 2015). The big challenge in this area was the management of EVD cases alerts; the active management of EVD was facing a lot of challenges such as: lack of good health care structures without standardized checklists of the cases, inadequate collaboration of certain families which hinder proper contact tracing activities. Thus, there were confirmed cases whose source cases were unknown. The mobility of cases was also a major challenge in managing this particular epidemic in Dubreka, patients with EVD could travel from a village to another, spreading the disease. It should be pointed out that our mission in Dubreka prefecture was characterized by lack of enthusiasm. Several times, the teams of supervisors, care teams and the Red Cross were assaulted by the villagers, making the task very difficult to all health workers.*"

Talking about Ebola outbreaks in the DRC, he also reported that: "*The management of Ebola in the DRC seems to be simplified by the facts that the population were a little informed about the disease, and rumors and legends seemed to disappear with time, because DRC has registered a high number of Ebola outbreaks and people are accepting to collaborate with health workers. The big challenges however have to do access the region concerned by the outbreak. These areas are often located in the Huge Equatorial Forest, which doesn't have good roads and Health structures*" (**Figures 4** and **5**).

Figure 4. Dr. Bona Ngoyi explaining hand washing techniques to the population of Boende (Equator province) during the 2014 Ebola outbreak in DRC.

Figure 5. Dr. Bona Ngoyi and his colleague using a boat to Likati (Equator province in DRC) to provide health care to the population during 2017 Ebola outbreak.

3.8. Treatment

There is no effective drug for EVD. Only supportive care could be administered, to sustain cardiac and renal functions with prudent use of perfusion. Oral rehydration can be recommended but sometimes not realistic because of throat pain, vomiting and intense fatigue [7].

In a clinical experiment conducted late in the 1995 Ebola outbreak in Kikwit, human convalescent blood was used for passive immunization to treat patients that had been infected naturally with ZEBOV; seven out of eight patients, who received blood transfusion from convalescent Ebola patients survived [72]. Such experiments, unfortunately, have not been repeated in further outbreaks because in vitro studies showed that antibodies against Ebola had no neutralizing activities. In addition, although monoclonal antibodies to the glycoprotein of Ebola virus showed protective and therapeutic properties in mice, they failed to protect NHP [73, 74]. Four laboratory workers in Russia who had possible Ebola exposure were treated with a combination of a goat-derived anti-Ebola immunoglobulin plus recombinant human interferon alfa-2. One of these patients had a high-risk exposure and developed clinical evidence of Ebola virus infection. All 4 patients recovered [75].

Many others Ebola vaccine candidates had been developed in the decade prior to 2014 [76]. In December 2016, Ebola virus disease was found to be 70–100% prevented by rVSV-ZEBOV vaccine, making it the first proven vaccine against the disease [77, 78].

4. Evolution of Ebola disease overtime

This section will focuses on key historical developments of Ebola disease over time.

4.1. Geographical evolution of the disease (country or regional spread)

For More than 3 decades (1976–2013), all major Ebola outbreaks were occurred in Central African countries: DRC, Uganda, Sudan, Congo, and Gabon. This could be linked to the Equatorial forest which covers all these countries: It has been shown that tropical rain forests of Africa to which the Western Congo Swamp Forests near Yambuku and Minkebé Forest in Gabon belong constitute a common ecosystem for Ebola virus emergence providing rich animal biodiversity and as such epidemics appear to be seasonal. Documented human and non-human EVD outbreaks occurred mainly during wet seasons, marked by fruit abundance. The index case of the 1995 EVD outbreak in Kikwit fell ill in January and the 1994 EVD outbreak among chimpanzees in the Tai forest occurred in November, at the end of the wet season [7]. It is also interesting to note that the center of outbreaks has always been in areas bordering on forests (ecotone forest-savannah in the Democratic Republic of Congo, savannah in Sudan) [79]. In Uganda, the regions (Luwero, Kibaale, Gulu) in which the outbreak occurs are areas bordering forests. The equatorial forest is a poorly developed region, where the population lives essentially by hunting [3]; this can increases contact with animals or animal's carcasses which could be potential reservoir of the virus.

Ebola outbreaks tend to occur more in a rural areas than urban areas, while the West Africa outbreak marks the first outbreak in a densely populated urban area within Conakry's large shanty towns [80]. Ebola outbreak has changed in its region of occurrence from central Africa to western African countries in 2013, and spread (isolated cases) all over the world (USA, UK, Italy). It is important to note that the index case in most of the outbreak comes from the rural area.

4.2. Severity evolution of the Ebola strain

Studies have shown that the high case-fatality rate for Ebola virus is attributed to Zaire Ebola virus species (50–90%), the case fatality for SUDV range from 40 to 60% [81, 82] and approximately 40% for BDBV [19]. Only one person has been infected with the Taï Forest strain and survived the illness [82, 83]. RESTV specie seems to be less pathogenic to humans. In a meta-analysis of WHO data from 20 outbreaks involving Zaire, Sudan and Bundibugyo Ebola species, including the 2014 West African outbreak, the average case fatality rate was estimated to be 65.4%, and ZEBOV case fatality was reported to decrease with time [84]. It is important to note that the more the country registered the outbreak, more the case fatality decrease: this could be explained by the fact that the health workers of a region which has registered several Ebola outbreaks will be trained to contain the disease than those in other regions which have never experienced the disease.

Another factor that could increase the severity of Ebola in Africa could be the co-infection of Ebola and Malaria or with other tropical diseases. The researchers found that malaria co-infection; extremes in age and delayed healthcare seeking behavior were all associated with mortality. Additionally, symptoms including disorientation, hiccups, diarrhea, and conjunctivitis, shortness of breath and muscle aches were all predictors of death in a very short time [85].

4.3. Changes and progress in diagnostic techniques for Ebola

Laboratory diagnosis of Ebola virus disease plays a critical role in outbreak response efforts; however, establishing safe and expeditious testing strategies for this high-biosafety-level pathogen in resource-poor environments remains extremely challenging. Since the discovery of Ebola virus in 1976 via traditional viral culture techniques and electron microscopy, diagnostic methodologies have trended towards faster, more accurate molecular assays. Importantly, technological advances have been paired with increasing efforts to support decentralized diagnostic testing capacity that can be deployed at or near the point of patient care [86]. Since the West African outbreak, efforts have been done to find a rapid and safe test for diagnosis of Ebola.

Diagnosis of Ebola has changed from cell culture, Antibody detection, Protein Antigen detection, conventional RT-PCR to Real-time RT-PCR. Real-time RT-PCR testing is an accurate and high-throughput modality and has become the standard for EVD diagnosis [86]. Current WHO guidelines recommend initial testing with an RDT when RT-PCR testing is not immediately available and to assist in triage and case management when clinical and laboratory resources are overwhelmed [87]. Furthermore, the requirement for collection and transport of venipuncture blood will continue to confer additional safety and logistical hurdles. In order to face these challenges, it is imperative that international partners work together with national health ministries to strengthen laboratory capacity in regions where Ebola is endemic, including the development of practical improvements to pre- and post-analytic processes and the training of local laboratory technicians in molecular diagnostic techniques, biosafety practices, and quality control [86].

5. Cost and effectiveness of Ebola outbreaks responses

The Ebola Response is highly complex. It requires the continuous effort by hundreds of different kinds of organizations and thousands of people to implement it quickly, effectively and efficiently [88]. Many countries all over the world have put public health measures (National response) in place to control EVD, apart from the supportive care that could be administrated to patient. These measures include checking and screening for EBOV at the airports and other points of entry, quarantine of people coming from regions associated with Ebola, and isolation of suspected and clinically diagnosed patients. The corner-stone for controlling an outbreak of EVD is to interrupt the viral transmission chain [7]. Management of survivors of EVD can also contributed to a good control of the outbreak. In the West Africa outbreak, Non-Conventional Humanitarian Interventions (NCHI) was declared as the principal strategy with major tasks at implementing relief logistics and the much-needed public health emergency responses to stamp out Ebola outbreak in vulnerable populations. The NCHI successfully supported operational containment efforts and lessons learnt in West Africa lay the foundation for accountable, transparent and innovative model for emergency response to global disease outbreaks in the most remote vulnerable populations [89]. In Uganda, Psychosocial Support (PSS) and community based volunteers in response to Ebola disease were introduced and the response was perceived to be very effective [90]. Other countries, which had experienced Ebola outbreaks before, opted to send their trained health workers to help those vulnerable regions to Ebola outbreak. For example, in August 2014 a team of 14 health workers from Uganda, which has "strong experience" of working with domestic Ebola outbreaks, had been deployed by the WHO to JFK Hospital in Monrovia, Liberia [91]. On 27 October 2014 it was announced that a further 30 Ugandan health workers were dispatched to affected countries in West Africa [92].

Various organizations around the world have always responded to all Ebola outbreaks: WHO, CDC, Medecins Sans Frontieres, etc. They work in collaboration or in association with health ministry of different countries which have been mapped out as areas with Ebola outbreaks. Special attention was taken to the West African outbreak. In August 2014, the outbreak was declared as an international public health emergency and a roadmap was published to guide and coordinate the international response to the outbreak, aiming to stop ongoing Ebola transmission worldwide within 6–9 months [93]. As of September 2014, a massive international response to the crisis was under way. The United Nations Mission for Ebola Emergency Response (UNMEER) had the task of overall planning and coordination, directing the efforts of the UN agencies, national governments, and other humanitarian actors to the areas where they are most needed [94]. UNMEER's objective was to work with others to stop the Ebola outbreak. UNMEER worked closely with governments, regional and international actors, such as the African Union (AU) and the Economic Community of West African States (ECOWAS), and with UN Member States, the private sector and civil society. Accra, in Ghana, served as a base for UNMEER, with teams in Guinea, Liberia and Sierra Leone [95, 96].

Funding is critical in responding to large and severe outbreaks of the nature of the West African Ebola outbreak. Many countries specifically donated to bring this health event under control. The US was the first country to donate to the Ebola response, then came the UK, Germany and the World Bank. The U.S. government allocated approximately $2.369 billion for Ebola response activities, including $798 million to CDC, $632 million to the Department of Defense, and $939 to the U.S. Agency for International Development. In addition to providing personnel, technical expertise, and resources to the response, these funds established three new emergency operations centers in Guinea, Liberia, and Sierra Leone [97]. Charity organizations, foundations and individuals also contributed financially to the global Ebola response.

6. Conclusion

EVD remains a global health problem. Identified in 1976 in Zaire (now Democratic Republic of Congo), Ebola is a highly contagious virus that manifests itself in the form of a hemorrhagic fever. The natural reservoir of the virus is fruit bat, which can contaminate humans directly or indirectly, through primates. Human-to-human transmission occurs through body fluids such as blood, stools, saliva, etc. The most severe Ebola outbreak began in December 2013 in south-eastern Guinea (West Africa) and extended to Liberia and Sierra Leone. The virus also affected Nigeria, Senegal, and Mali and even beyond the African continent (USA, UK, and Italy); at the end of October 2014, there were nearly 5000 deaths caused by the EVD. The latest outbreak ended on 2 July 2017, in the Likati Health Zone in Bas-Uélé Province of the Democratic Republic of the Congo (DRC).

Aside the high CFR associated with EVD (between 25 and 90%), a worrying phenomenon has been the continuous loss of already inadequate critical clinical and support workforce to the disease in outbreak and response settings. In low resource settings, the challenge has been the inability of the ministries of health to provide the adequate medical consumables necessary to protect the health workers and to ensure proper infection control practices. While health staff in these regions are gaining more knowledge and experience in dealing with occasional outbreaks, these logistical challenges considerably hinder their practice and further expose them to infections.

Though there is no effectively established drug for the treatment of EVD, recent advancement in vaccine development present a ray of hope that EVD could potentially be a vaccine preventable disease.

It is noted that Ebola outbreaks have the potential to escalate in resource-challenged regions with non-existent or very basic health infrastructure, poor road networks making the communities hard-to-reach and the primary co-existence or apparent contact with reservoirs. In the instance of the 2014 West African outbreak which was reported in densely populated urban centers, the disease spread rapidly due to weak health systems, inadequate infection prevention and control measures and non-responsive disease surveillance systems. Partnership between international organizations and ministries of health of Ebola endemic countries therefore become crucial to prevent, detect and appropriately respond to surges of Ebola among populations. This may be done through direct support to strengthen disease surveillance systems to ensure total coverage of all regions/provinces and districts, strengthening event based surveillance to establish early warning systems for disease outbreaks, building the capacity of local laboratories and encouraging the formation of a network of laboratories within and among neighboring countries while prioritizing infection prevention and control measures. Considering the ease of global spread of this disease in light of the rapid migratory patterns in recent years, the burden of preventing and controlling this disease rests on the Public health authorities of all countries over the world and their partners to work towards:

- Organizing community education campaigns designed to give more details on the viral nature of the EVD.

- Increase awareness through health education of the population through campaigns about EVD with particular attention to: hygienic measures, cooking of bush meat as long as possible, avoiding coming into contact with the biological fluids from persons suspected or diagnosed with a hemorrhagic fever.
- Communities affected by the Ebola virus must inform the population of the measures taken to contain the outbreak, including safe, dignified burial and funeral practices. People who have died from this infection must be buried quickly and without excessive risk to those who carry out the burial.
- Inform the population about the physio-pathological aspect of the virus in order to reduce rumors and false beliefs about the disease.
- Expanding training of qualified people for better management of the outbreak, and increase supply of medical materials to isolated rural areas.
- Establishing structures for early detection of any future outbreaks. Motivating the health care professionals, especially those working in the zone with previous Ebola outbreaks.
- For travelers, it is important to impose quarantine to any person suspected or diagnosed with EVD.
- Laboratory research should be carried out in biosafety. Procedures on sterilization and decontamination must be rigorously applied to avoid laboratory contamination.

Author details

Kasangye Kangoy Aurelie[1,2*], Mutangala Muloye Guy[3,4], Ngoyi Fuamba Bona[5], Kaya Mulumbati Charles[6], Avevor Patrick Mawupemor[7] and Li Shixue[1]

*Address all correspondence to: aureliekasangye@yahoo.fr

1 School of Public Health, Social Medicine and Health Management Department, Shandong University, Jinan, Shandong, China

2 School of Public Health, University of Lubumbashi, Lubumbashi, Democratic Republic of the Congo

3 School of Medicine, Department of Obstetrics and Gynecology, Qilu Hospital, Shandong University, Jinan, Shandong, China

4 School of Medicine, Department of Obstetrics and Gynecology, University of Lubumbashi, Lubumbashi, Democratic Republic of the Congo

5 Public Health Ministry of the Democratic Republic of the Congo, Department of Epidemiology, Kinshasa, Democratic Republic of the Congo

6 School of Medicine, Department of Public Health, University of Lubumbashi, Lubumbashi, Democratic Republic of the Congo

7 School of Medical and Health Sciences, Mountcrest University College, Accra, Ghana

References

[1] WHO. Ebola Virus Disease. WHO Africa. http://www.afro.who.int/health-topics/ebola-virus-disease, 2017 WHO/Regional office for Africa

[2] Hewlett BL, Hewlett BS. Providing care and facing death: Nursing during Ebola outbreaks in Central Africa. Journal of Transcultural Nursing. 2005;**16**:2891

[3] Kangoy AK, Muloye GM, Avevor PM, Shixue L. Review of past and present Ebola hemorrhagic fever in the Democratic Republic of Congo 1976-2014. African Journal of Infectious Diseases. 2016;**10**(1):38-42

[4] Le Point international. Le découvreur belge de l'Ebola ne craint pas une épidémie majeure hors d'Afrique. http://www.lepoint.fr/monde/

[5] le monde.fr.1976, à la découverte du virus Ebola. 2014. www.lemonde.fr/planete/article/

[6] Torpiano P, Pace D. Ebola: Too far or so close? Malta Medical Journal. 2014;**26**:32-40

[7] Muyembe JJ, Mulangu S, Masumu J, Kayembe JM, et al. Ebola virus outbreaks in Africa: Past and present. The Onderstepoort Journal of Veterinary Research. 2012;**79**:1-8

[8] Netesov SV, Feldmann H, Jahrling PB, Klenk HD, Sanchez. Family Filoviridae. In: Virus Taxonomy—Seventh Report of the International Committee on Taxonomy of Viruses. San Diego, USA: Academic Press; 2000. pp. 539-548. ISBN 0-12-370200-3

[9] Pringle CR. Virus taxonomy-San Diego 1998. Archives of Virology. 1998;**143**(7):1449-1459

[10] Feldmann H, Geisbert TW, Jahrling PB, Klenk H-D, et al. Family Filoviridae. In: Fauquet CM, editor. Eighth Report of the International Committee on Taxonomy of Viruses. San Diego, USA: Elsevier/Academic Press; 2005. pp. 645-653. ISBN 0-12-370200-3

[11] Mayo MA. ICTV at the Paris ICV: Results of the plenary session and the binomial ballot. Archives of Virology. 2002;**147**(11):2254-2260

[12] Kuhn JH, Becker S, Ebihara H, Geisbert TW, et al. Proposal for a revised taxonomy of the family Filoviridae: Classification, names of taxa and viruses, and virus abbreviations. Archives of Virology. 2010;**155**(12):2083-2103. PMC 3074192 Freely accessible

[13] International Committee on Taxonomy of Viruses. Virus Taxonomy: 2013 Release. https://talk.ictvonline.org/taxonomy/

[14] Bowen ETW, Lloyd G, Harris WJ, Platt GS, et al. Viral haemorrhagic fever in southern Sudan and northern Zaire. Preliminary studies on the aetiological agent. Lancet. 1977;**309**(8011):571-573. PMID 65662

[15] http://whqlibdoc.who.int/bulletin/1978/Vol56-No2/bulletin_1978_56(2)_247-270.pdf

[16] Le Guenno B, Formenty P, Wyers M, Gounon P, Walker F, Boesch C. Isolation and partial characterisation of a new strain of Ebola virus. Lancet. 1995;**345**(8960):1271-1274. PMID 7746057

[17] Kuhn JH. Filoviruses. A compendium of 40 years of epidemiological, clinical, and laboratory studies. Archives of Virology. Supplementum. 2008;**20**:13-360. PMID 18637412

[18] Wamala JF, Lukwago L, Malimbo M, Nguku P, Yoti Z, et al. Ebola hemorrhagic fever associated with novel virus strain, Uganda, 2007-2008. Emerging Infectious Diseases. 2010;**16**(7):1087-1092. PMC 3321896 Freely accessible. PMID 20587179

[19] MacNeil A, Farnon EC, Wamala J, Okware S, et al. Proportion of deaths and clinical features in Bundibugyo Ebola virus infection, Uganda. Emerging Infectious Diseases. 2010;**16**(12):1969-1972. PMC 3294552 Freely accessible. PMID 21122234

[20] Geisbert TW, Jahrling PB. Use of immunoelectron microscopy to show Ebola virus during the 1989 United States epizootic. Journal of Clinical Pathology. 1990;**43**(10):813-816. PMC 502829 Freely accessible. PMID 2229429

[21] Leroy EM, Rouquet P, Formenty P, Souquière SF, Kilbourn A, Froment JM. Multiple Ebola virus transmission events and rapid decline of central African wildlife. Science. 2004;**303**:387-390

[22] Leroy EM, Kumulungui B, Pourrut X, Rouquet P, Hassanin A, Yaba P, Délicat A, Paweska JT, Gonzalez JP, Swanepoel R. Fruit bats as reservoirs of Ebola virus. Nature. 2005;**438**:575-576

[23] Pourrut X, Kumulungui B, Wittmann T, Moussavou G, et al. The natural history of Ebola virus in Africa. Microbes and Infection. 2005;**7**(7-8):1005-1014

[24] Morvan JM, Deubel V, Gounon P, Nakouné E, et al. Identification of Ebola virus sequences present as RNA or DNA in organs of terrestrial small mammals of the Central African Republic. Microbes and Infection. 1999;**1**(14):1193-1201

[25] Groseth A, Feldmann H, Strong JE. The ecology of Ebola virus. Trends in Microbiology. 2007;**15**(9):408-416

[26] Olival KJ, Islam A, Yu M, Anthony SJ, et al. Ebola virus antibodies in fruit bats, Bangladesh. Emerging Infectious Diseases. 2013;**19**(2):270-273

[27] CDC. Transmission of Ebola Virus; 2015. https://www.cdc.gov/vhf/ebola/transmission/index.html

[28] Biek R, Walsh PD, Leroy EM, Real LA. Recent common ancestry of Ebola Zaire virus found in a bat reservoir. PLoS Pathogens. October 2006;**2**(10):e90, 0885-0886

[29] Organisation Mondiale de la Santé (OMS). Ce que l'on sait à propos de la transmission interhumaine du virus Ebola : Évaluation de la situation; 2014. http://www.who.int/mediacentre/news/ebola/06-october-2014/fr/

[30] WHO. Ebola Virus Disease. Fact Sheet No 103. Disease Fact Sheets [Internet]. 2014. http://www.who.int/mediacentre/factsheets/fs103/en

[31] Rambaut A. Case fatality rate for ebolavirus. Epidemic; 2014. http://epidemic.bio.ed.ac.uk/ebolavirus_fatality_rate

[32] Walsh PD, Abernethy KA, Bermejo M, Beyers R, De Wachter P, et al. Catastrophic ape decline in western equatorial Africa. Nature. 2003;**422**:611-614

[33] Ebola Situation Report (PDF). World Health Organization. 21 January 2015. Retrieved 22 January 2015. http://apps.who.int/iris/bitstream/10665/149314/1/roadmapsitrep_21Jan2015_eng.pdf?ua=1

[34] Reuters. Mali Confirms New Case of Ebola, Locks Down Bamako Clinic. 12 November 2014. Retrieved 15 November 2014. https://in.reuters.com/article/health-ebola-mali/update-1-mali-confirms-new-case-of-ebola-locks-down-bamako-clinic-idINL6N0T15CN20141112

[35] WHO. Ebola Response Roadmap Situation Report Update. World Health Organization. 7 November 2014. Retrieved 7 November 2014. http://apps.who.int/iris/bitstream/10665/137592/1/roadmapsitrep_7Nov2014_eng.pdf?ua=1

[36] WHO. Ebola Response Roadmap – Situation Report. World Health organization. 31 December 2014. Retrieved 1 January 2015. The reported case fatality rate in the three intense-transmission countries among all cases for whom a definitive outcome is known is 71%. http://apps.who.int/ebolaweb/sitreps/20141231/20141231.pdf

[37] WHO. Ebola Virus Disease – Italy. WHO. http://www.who.int/csr/don/13-may-2015-ebola/en/

[38] Una enfermera que atendió al misionero fallecido García Viejo, contagiada de ébola (in Spanish). El Mundo. 6 October 2014. Retrieved 6 October 2014. http://www.elmundo.es/madrid/2014/10/06/5432bb62e2704e347a8b4577.html?a=a733fe5654cd56f5e1d50d769a9b4204&t=1412616936

[39] Ebola Outbreak Situation Report (PDF). WHO. 8 October 2014. Retrieved 15 October 2014. http://apps.who.int/iris/bitstream/10665/136020/1/roadmapsitrep_8Oct2014_eng.pdf?ua=1

[40] Ebola Virus Epidemic Situation Map, Simplified. 2014. https://en.wikipedia.org/wiki/West_African_Ebola_virus_epidemic#/media/File:2014_ebola_virus_epidemic_in_West_Africa_simplified.svg

[41] WHO Ebola Response Team. Ebola virus disease in West Africa – The first 9 months of the epidemic and forward projections. New England Journal of Medicine. 23 September 2014;**371**:1481-1495

[42] West Africa Outbreak. Centers for Disease Control and Prevention. 2014. Retrieved 11 April 2015. https://www.cdc.gov/vhf/ebola/outbreaks/2014-west-africa/previous-updates.html

[43] WHO Director-General Briefs Media on Outcome of Ebola Emergency Committee. World Health Organization. Retrieved 2 April 2016. http://www.who.int/csr/disease/ebola/guinea-flareup-update/en/

[44] WHO. Interim Advice on the Sexual Transmission of the Ebola Virus Disease. World Health Organization. Retrieved 28 October 2016. http://www.who.int/reproductivehealth/topics/rtis/ebola-virus-semen/en/

[45] WHO Director-General Addresses the Executive Board. World Health Organization. Retrieved 27 January 2016. http://www.who.int/csr/disease/ebola/response/en/

[46] Ebola Data and Statistics. World Health Organisation. Retrieved 9 June 2016. http://apps.who.int/gho/data/view.ebola-sitrep.ebola-summary-latest?lang=en

[47] Ebola Situation Report. Ebola Data and Statistics. World Health Organization. 12 January 2015. Retrieved 28 January 2015. http://apps.who.int/gho/data/view.ebola-sitrep.ebola-summary-20150112?lang=en

[48] Identification of Ebola Virus on Firefly Dx. http://psidcorp.com/system/files/media/GenArraytion%20Ebola%20Assay%20on%20Firefly%20Dx.pdf

[49] WHO Declares Nigeria Ebola-Free. WHO. Retrieved 22 July 2016. http://www.afro.who.int/media-centre/news

[50] Diseases – Activity 1 – Glossary, page 3 of 5. science.education.nih.gov. Retrieved 2010-11-03

[51] WordNet Search – 3.0. Princeton University. wordnetweb.princeton.edu. Retrieved 3 November 2010. http://wordnetweb.princeton.edu/perl/webwn?s=index%20case

[52] Sporadic STEC O157 Infection: Secondary Household Transmission in Wales. USA: Centers for Disease Control and Prevention. www.cdc.gov. 1 January 1994. Retrieved 3 November 2010

[53] WHO. Ebola hemorrhagic fever in Zaire. Bulletin of the World Health Organization. 1979;**56**:271-293

[54] Leroy EM, Epelboin A, Mondonge V, Pourrut X, et al. Human Ebola outbreak resulting from direct exposure to fruit bats in Luebo, Democratic Republic of Congo. Vector-Borne and Zoonotic Diseases. 2007;**9**:723-728

[55] Georges AJ, Leroy EM, Renaud AA, et al. Ebola hemorrhagic fever outbreaks in Gabon, 1994-1997: Epidemiologic and health control issues. Journal of Infectious Diseases. 1999;**179**:S65-S75

[56] WHO. Ebola haemorrhagic fever – South Africa [469 KB, 8 pages]. Weekly Epidemiological Record. 1996;**71**(47):359

[57] Okware S, Omaswa FG, Zaramba S, Opio A, Lutwama JJ, Kamugisha J, Rwaguma EB, Kagwa P, Lamunu M. An outbreak of Ebola in Uganda. Tropical Medicine & International Health. 2002 Dec;**7**(12):1068

[58] Bwaka MA, Bonnet MJ, Calain P, Colebunders R, De Roo A, Guimard, et al. Organization of patient care during the Ebola hemorrhagic fever epidemic in Kikwit, Democratic Republic of the Congo, 1995. Journal of Infectious Diseases. 1 February 1999; **179**(Supplement_1):S268-S273

[59] Okware S. Managing Ebola in low-resource settings: Experiences from Uganda. Medicine, Infectious Diseases, Ebola; August 2016. https://www.intechopen.com/books/ebola/managing-ebola-in-low-resource-settings-experiences-from-uganda

[60] Finding Ebola's 'patient zero'. The Guardian. Retrieved 28 November 2014

[61] Hollow tree in Guinea was Ebola's Ground Zero, scientists say. Mail & Guardian Africa. AFP. 30 December 2014. Retrieved 1 January 2016

[62] Grady D, Fink S. Tracing Ebola's breakout to African 2-year-old. The New York Times. Retrieved 11 April 2015

[63] Stylianou N. How world's worst Ebola outbreak began with one boy's death. BBC News. 27 November 2014. Retrieved 11 April 2015

[64] Maganga D, Kapetshi J, Berthet N, Kebela Ilunga B, et al. Ebola virus disease in the Democratic Republic of Congo. The New England Journal of Medicine. 2014;**371**:2083-2091

[65] WHO. External Situation Report 1. Regional Office for Africa. Ebola Virus Disease–Democratic Republic of the Congo. World Health Organization. 15 May 2017. Retrieved 16 May 2017. http://apps.who.int/iris/handle/10665/255713?locale=en

[66] Garrett L. The Coming Plague: Newly Emerging Diseases in a World Out of Balance. London: Virago Press; 1994. pp. 100-105. ISBN 1-86049-211-8

[67] WHO. Health Worker Ebola Infections in Guinea, Liberia and Sierra Leone. May 2015. http://www.who.int/csr/resources/publications/ebola/health-worker-infections/en/

[68] Grinnell M, Dixon MG, Patton M, Fitter D, et al. Ebola virus disease in health care workers. CDC Morbidity and Mortality Weekly Report. October 2, 2015;**64**(38):1083-1087

[69] Hero nurse Pauline Cafferkey could have contracted deadly Ebola at Christmas Day service. The Telegraph. 30 December 2014

[70] Grady D. Ebola is Diagnosed in Texas, First Case Found in the U.S. The New York Times/Health. Sept. 30, 2014. https://www.nytimes.com/2014/10/01/health/airline-passenger-with-ebola-is-under-treatment-in-dallas.html

[71] Gidda M. Fear and Rumors Fueling the Spread of Ebola. Time Health 2014 / Infectious disease. http://time.com/3092855/ebola-fear-rumors/

[72] Mupapa K, Massamba M, Kibadi K, Kuvula K, et al. Treatment of Ebola hemorrhagic fever with blood transfusions from convalescent patient. Journal of Infectious Diseases. 1999;**179**(Supplement 1):S18-S23

[73] Gupta M, Mahanty S, Bray M, Ahmed R, Rollin PE. Passive transfer of antibodies protects immunocompetent and imunodeficient mice against lethal Ebola virus infection without complete inhibition of viral replication. Journal of Virology. 2001;**75**:4649-4654. PMID:11312335

[74] Oswald WB, Geisbert TW, Davis KJ, Geisbert JB, et al. Neutralizing antibody fails to impact the course of Ebola virus infection in monkeys. PLoS Pathogens. January 2007;**3**(1):e9, 0062-0066

[75] Ebola Virus Infection Treatment & Management. Updated: Jun 03, 2016. http://emedicine.medscape.com/article/216288-treatment#d9

[76] Richardson JS, Dekker JD, Croyle MA, Kobinger GP. Recent advances in Ebolavirus vaccine development. Human Vaccines. 2010;**6**(6):439-449

[77] Henao-Restrepo AM et al. Efficacy and effectiveness of an rVSV-vectored vaccine in preventing Ebola virus disease: Final results from the Guinea ring vaccination, open-label, cluster-randomised trial (Ebola Ça Suffit!). The Lancet. 2016;**389**(10068):505-518. PMID 28017403. Retrieved 27 December 2016

[78] Berlinger J. Ebola vaccine gives 100% protection, study finds. CNN. Retrieved 27 December 2016

[79] Morvan JM, Nakouné E, Deubel V, Colyn M. Forest ecosystems and Ebola virus. Bulletin de la Société de Pathologie Exotique. 2000;**93**(3):172-175

[80] Diallo B. Ebola en Guinée: l'ONG Plan-Guinée craint une aggravation de l'épidémie. 2014. Africaguinée.com. Available from: http://africaguinee.com/articles/2014/03/30/ebola-en-guinee-l-ong-plan-guinee-craint-une-aggravation-de-l-epidemie

[81] Feldmann HK, Geisbert TW. Ebola haemorrhagic fever. The Lancet. 2011;**377**(9768):849-862

[82] Leroy EM, Gonzalez JP, Baize S. Ebola and Marburg haemorrhagic fever viruses: Major scientific advances, but a relatively minor public health threat for Africa. Clinical Microbiology and Infection. 2011;**17**(7):964

[83] de Wit E, Feldmann H, Munster VJ. Tackling Ebola: New insights into prophylactic and therapeutic intervention strategies. Genome Medicine. 2011;**3**(1):5

[84] Lefebvre A, Fiet C, Belpois-Duchamp C, Tiv M, Astruc K, Aho Glele LS. Case fatality rates of Ebolavirus diseases: A meta-analysis of World Health Organization data. Médecine et Maladies Infectieuses. 2014;**44**(9):412

[85] Medical Xpress. New scoring system predicts Ebola severity. February 2, 2017. https://medicalxpress.com/news/2017-02-scoring-ebola-severity

[86] Broadhurst MJ, Brooks TJG, Pollock NR. Diagnosis of Ebola virus disease: Past, present, and future. Clinical Microbiology Reviews. 1 October 2016;**29**(4):773-793

[87] World Health Organization. Interim Guidance on the Use of Rapid Ebola Antigen Detection Tests. Geneva, Switzerland: World Health Organization; 2015. http://www.who.int/csr/resources/publications/ebola/ebola-antigen-detection/en/

[88] Global Response to Ebola. https://ebolaresponse.un.org/

[89] Talla M, Chidiebere UE, Olalubi OA, Wurie I, Jonhson JK, Ngogang JY, Tambo E. Effectiveness of non-conventional humanitarian responses on Ebola outbreak crisis in West Africa. Journal of Pharmacovigilance. 2015;**3**(4):98

[90] Thormar SB. Evaluation of Ebola Response-Uganda 2012. Feb 2013. www.adore.ifrc.org/Download.aspx?FileId=42478&.pdf

[91] WHO. Ugandan Team Brings Ebola Experience to Liberia. World Health Organization. August 2014. http://www.who.int/features/2014/ugandan-ebola-team/en/

[92] Uganda sends more medical personnel to Ebola-hit West Africa. Shanghai Daily. 27 October 2014. http://www.shanghaidaily.com/article/article_xinhua.aspx?id=249178

[93] Ebola 'threat to world security'- UN Security Council. BBC News. 18 September 2014. Retrieved 18 September 2014

[94] WHO warns Ebola response could cost $1B. The Hill. 16 September 2014. Retrieved 16 September 2014

[95] UN Mission for Ebola Emergency Response (UNMEER). United Nations. Retrieved 4 October 2014. http://ebolaresponse.un.org/un-mission-ebola-emergency-response-unmeer

[96] WHO welcomes decision to establish United Nations Mission for Ebola Emergency Response. Time. EIN Presswire. 19 September 2014. https://www.einnews.com/pr_news/224890550/who-welcomes-decision-to-establish-united-nations-mission-for-ebola-emergency-response

[97] CDC. Cost of the Ebola Epidemic. https://www.cdc.gov/vhf/ebola/pdf/impact-ebola-economy.pdf

Chapter 2

Registering Ebola Virus Disease (EVD) Both as a Multiple Cause of Death and as a Notifiable Disease in Africa: Comparison Between the Ideal and the Reality

Sulaiman Bah

Additional information is available at the end of the chapter

http://dx.doi.org/10.5772/intechopen.72593

Abstract

The chapter explores the possibility of registering Ebola virus disease (EVD) as a multiple cause of death (part of the civil registration/vital statistics (CR/VS) system) in addition to being a notifiable disease (part of the disease surveillance system). The linkage between the two systems is established, followed by a framework showing how each of the systems would work in the ideal situation. A scoring system is developed and used to score each dimension of this ideal system, giving a total score of 23. This tool can be used to assess the extent to which the EVD is registered both as a multiple cause of death and as a notifiable disease in Africa. The application of the tool requires that the Ebola virus disease is coded at the fourth digit ICD-10 level and that multiple causes of death are routinely collected in the first place. The country that is closest to satisfying these criteria is South Africa. The application of the tool to South Africa data showed that South African system was "fair" (between "poor" and "good"). The results are shown, discussed and recommendations are made for improving two systems in Africa.

Keywords: Ebola virus disease, disease surveillance system, civil registration/vital statistics system, multiple causes of death, ICD-10

1. Introduction

The unexpected outbreak of the Ebola virus disease (EVD) in West Africa during 2014–2015 sadly led to over 10,000 deaths [1]. The resulting amount and diversity of EVD-related research that followed was impressive. Some tried to produce estimates of EVD cases and related deaths

[2], some tried to better understand the epidemiology of EVD [3], while others tried to analyze some of the structural factors that led to the disaster [4]. Analytical reports on the West African EVD epidemic (including a CDC report) often mention that there was no prior EVD outbreak in the main affected countries of Sierra Leone, Guinea and Liberia [5]. While this observation is true, what it fails to mention is that the region bordering the three countries was already endemic to Lassa fever, another viral hemorrhagic disease. According to published findings, Lassa fever had been detected in that region as early as the 1970s. A Lassa fever outbreak had occurred in Liberia in 1972, and in a hospital-based study in Liberia in 1976–1977, Lassa fever antibodies were found to be present in 8.4% of the 844 sera specimen studied [6]. In a serological survey carried out in Liberia in 1978–1979, it was found that of 433 sera specimen studied, 16% tested positive for Lassa fever, 6% for Ebola virus and 1% for Marburg virus [7]. The latter study concluded the following: 'the results seem to indicate that Liberia has to be included in the Ebola and Marburg virus endemic zones' [7]. Other studies subsequently confirmed the endemicity of Lassa fever in both Guinea and Sierra Leone [8]. In short, the 2014/2015 Ebola epidemic in West Africa had been preceded by decades of the endemicity of other hemorrhagic fevers in the region. This suggests that EVD may lay hidden for many years before it breaks out as an epidemic. Hence, in addition to EVD being a notifiable disease, it would make sense to search for EVD among other causes present at death, in other words, as a multiple cause of death. This would strengthen the monitoring system for long-term prediction of possible outbreaks of EVD.

The rest of the section discusses the following topics: linkage between disease surveillance system and multiple causes of death system; the setup for an ideal disease surveillance system and the setup for a practical and efficient system for collecting data on multiple causes of death. The findings of these sections are used in developing the methods section which follows. The results are presented and discussed. Subsequently, the chapter ends with some concluding remarks.

1.1. Linkage between disease surveillance system and multiple causes of death system

Figure 1 shows a simplified relationship between the disease surveillance system and the system for producing multiple causes of death statistics. When someone contracts a notifiable disease, this may or may not result in contact with the healthcare delivery system. In an ideal system, once the patient with the notifiable disease gets in contact with the healthcare delivery system, the case is notified to the authorities and the details are entered into the disease surveillance system and the necessary public health action is taken. Even if this patient does not contact the healthcare delivery system, it is possible to enter the information in the disease surveillance system via lay reporting. After the patient dies (in or out of hospital), it is only when the death is reported that it becomes a part of the civil registration/vital statistics (CR/VS) system. In the processing of the cause of death, if underlying cause of death coding is used, the notifiable disease may go unreported if it is not the underlying cause of death. Hence, it is only through multiple cause of death coding that the notifiable disease (which is

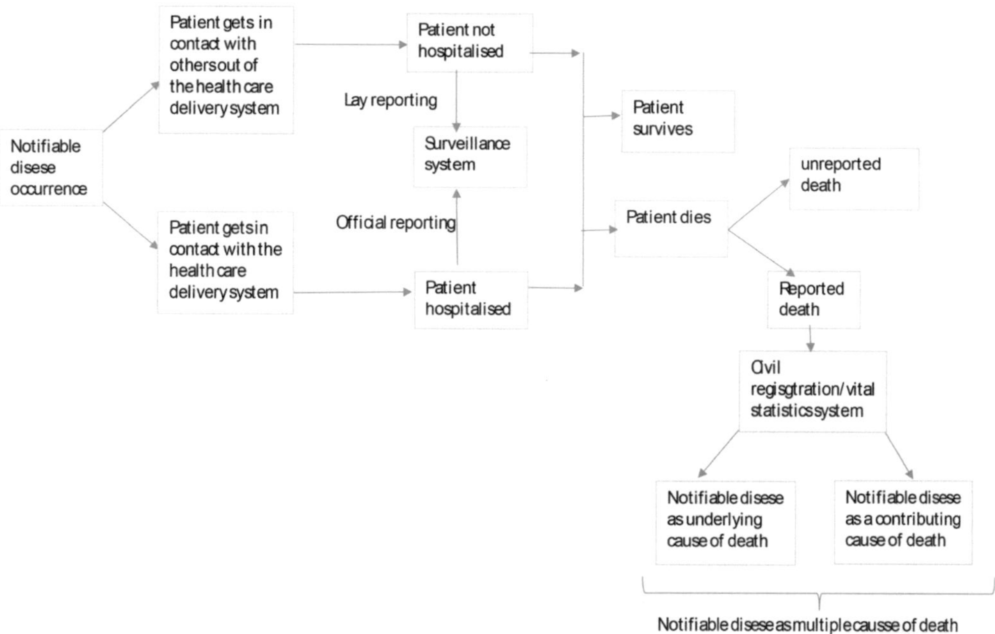

Figure 1. Linkage between disease surveillance system and multiple causes of death system.

not the underlying cause of death) present at death would appear in official vital statistics. The implication of this schema is that for notifiable diseases in general, the expectation is that the number of cases reported in the surveillance system should be more than the number of deaths with the notifiable disease as a multiple cause of death. However, for diseases with very high fatality rate and situations in which the surveillance systems and the vital registration systems are working very well, the two figures would be close to each other. The wider the difference between the data coming from the two systems, the more departure from the ideal, for either of the systems or both systems.

1.2. The setup for an ideal disease surveillance system

Figure 2 shows the framework for an ideal disease surveillance system in developing countries. While disease surveillance systems are in place in most countries, their efficiency varies markedly. The reasons for the marked variation are many. The first is the scope of entities included in the disease notification system. For many, the emphasis is mostly on government hospitals, laboratories and clinics. Private entities are either not properly integrated or not given due importance, as was found in a 2013 study in Iran [9]. The second is the complication or perceived complication of the notification process. The more complicated the process is, and the less incentives there are, the lower the reporting of notifiable diseases. The third is the lack of penalty (or low penalty) for failing to report. All of these factors contribute toward low reporting rates for notifiable diseases.

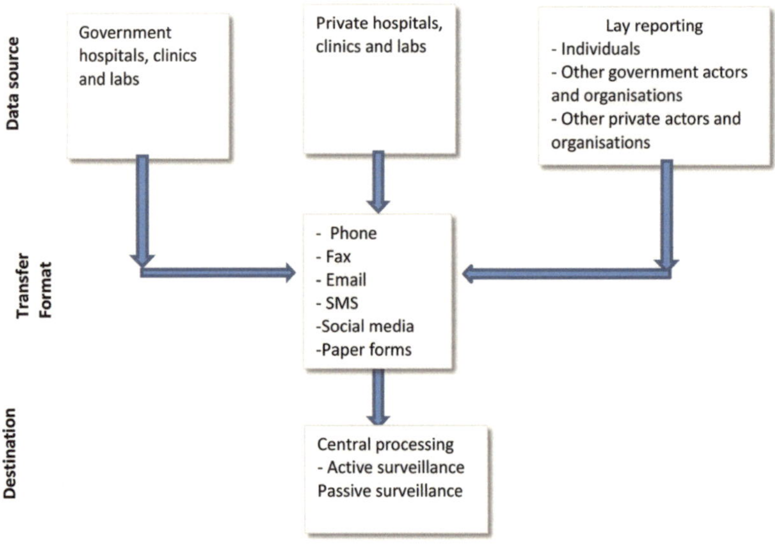

Figure 2. Framework for an ideal disease surveillance system in developing countries.

1.3. The setup for a practical and efficient system for collecting data on multiple causes of death

Unlike disease surveillance systems, there is no 'ideal system' for collecting data on multiple causes of death. This is because the vital statistics systems are a by-product of the civil registration system, which has diverse arrangements in different countries. A CR/VS system that produces the desired outcome is an acceptable system, irrespective of the arrangements used. The desired outcome of acceptable system for producing statistics on multiple causes of death is as follows:

1. Timely production of statistics, not more than 1 year after the end of the reference year
2. Regular official tabulation of multiple causes of death tables
3. The use of software for the automatic selection of underlying causes of death
4. The use of ICD-10 (at the four digit level) for coding causes of death
5. The negligible proportion of 'ill-defined causes of death'

The fourth point is mentioned because EVD is only properly coded at the four-digit level in ICD-10 (A98.4).

In passing, it is worth mentioning that the country that best satisfies all these criteria is arguably Australia. Australia regularly publishes official tables on multiple causes of death. It uses ICD-10, at the fourth digit level, to code causes of death. It uses IRIS software for the automatic selection of underlying causes of death. Its proportion of ill-defined causes of death is negligible. Lastly, the 2016 statistics on multiple causes of death was published in 2017. As such, the Australian system for producing multiple causes of death is the closest to the ideal.

2. Methods

2.1. Criteria for comparing surveillance and multiple causes of death systems against the ideal

The features of the ideal systems described above have been converted into a table (**Table 1**) with scores given for the different dimensions. Each dimension is scored as either binary (0 or 1) or ordinal (from 0 to 3). Using this scoring system, the maximum score (the ideal system) is 23. Any system scoring 18 or over can be rated as good. Any system scoring between 13 and 17 can be rated as fair, and any system rating below 13 can be rated as poor. Information on dimensions from 1.1 to 3.1 can be obtained from the literature, by studying the working of the surveillance system and the multiple causes of death system. The dimension 3.2, however, could only be assessed through analysis of data, through the direct comparison of the number of cases from the surveillance system and the number of deaths from the multiple causes of death system for the same year and the same cause.

No.	Dimension	Scoring system (Binary or ordinal)
1.	**Disease surveillance system**	
1.1	Collections of data from government hospitals, clinics and laboratories	0-Absent 1-Low 2-Moderate 3-High
1.2	Collections of data from private hospitals, clinics and laboratories	0-Absent 1-Low 2-Moderate 3-High
1.3	Use of lay reporting	0-Absent 1-Present
1.4	Multiple transfer format for reporting	0-None 1-One format 2-Two formats 3-Three or more formats
1.5	Central processing	0-Absent 1-Present
1.6	Using of active surveillance for highly infectious disease	0-Not used 1-Used
1.7	ICD-10 coding	0-Not used 1-Used
1.8	Use of four-digit coding	0-Not used 1-Used

No.	Dimension	Scoring system (Binary or ordinal)
2.	**Multiple causes of death system**	
2.1	Timeliness of reporting of causes of death	1-After 4 or more years 2-Within 3 years 3-Within 2 years 4-Within 1 year
2.2	Regular official tabulation for multiple causes of death	0-Absent 1-Present
2.3	Software for automating the coding of causes of death and the selection of underlying cause of death	0-Absent 1-Present
2.4	ICD-10 coding	0-Absent 1-Present
2.5	Use of four-digit coding	0-Absent 1-Present
3.	**Overlap between the surveillance system and the multiple causes of death system**	
3.1	Official linkage between the surveillance system and the multiple causes of death system	0-Absent 1-Present
3.2	The number of reported cases in the surveillance system being equal to, or more than the reported multiple causes of death for the same disease for the same reporting year	0-No 1-Yes

Table 1. Criteria for comparing surveillance and multiple causes of death systems against the ideal.

3. Results

The application of the criteria to Africa starts with some inclusion criteria. Since we are discussing EVD and multiple causes of death, the main inclusion criteria for African countries to be included in this study are two: the official collection and publication of data on multiple causes of death and the use of ICD-10 coding. According to the data included in the global health data exchange (GHDx), the only African countries submitting mortality data to WHO using ICD-10 coding are as follows: Egypt, Morocco, Tunisia, Cape Verde, Zambia, Mauritius, Seychelles and South Africa. Of these countries, only South Africa collects and publishes data on multiple causes of death.

The South African national statistics office, Statistics South Africa (Stats SA), routinely collects, analyses and publishes data on multiple causes of death. It has been routinely publishing data on multiple causes of death, starting from the 1997 data. The causes of death coding were initially done at the three-digit level, but recently, it has moved on to four-digit coding. South Africa also has decades-old functional disease notification system for reporting and analyzing notifiable diseases (the South African Institute for Medical Research was established in 1912). The number of notifiable diseases in South Africa is over 40 and includes Crimean-Congo hemorrhagic fever (CCHF) (ICD-10: A98.0) and 'other hemorrhagic fevers of Africa,' which

No.	Dimension	Score	Source/note
1.	**Disease surveillance system**		
1.1	Collections of data from government hospitals, clinics and laboratories	2	The surveillance system has been independently assessed and rated as being 64% complete [11]
1.2	Collections of data from private hospitals, clinics and laboratories	1	'There are no legal provisions for laboratories to notify communicable diseases.' [11]
1.3	Use of lay reporting	0	Not mentioned in the reporting mechanism
1.4	Multiple transfer format for reporting	2	'The NDSS in South Africa is a paper-based system...'[11] 'All suspected VHF cases require an immediate telephonic notification' [12]
1.5	Central processing	1	The National Institute for Communicable Diseases (NICD) is responsible for disease surveillance
1.6	Using of active surveillance for highly infectious disease	1	
1.7	ICD-10 coding	0	The description used 'other hemorrhagic fevers of Africa' in disease surveillance is not part of the ICD-10 description
1.8	Use of four-digit coding	0	
2.	**Multiple causes of death system**		
2.1	Timeliness of reporting of causes of death	2	[13]
2.2	Regular official tabulation for multiple causes of death	1	
2.3	Software for automating the coding of causes of death and the selection of underlying cause of death	1	
2.4	ICD-10 coding	1	
2.5	Use of four-digit coding	1	
3.	**Overlap between the surveillance system and the multiple causes of death system**		
3.1	Official linkage between the surveillance system and the multiple causes of death system	0	Not mentioned
3.2	The number of reported cases in the surveillance system being equal to, or more than the reported multiple causes of death for the same disease for the same reporting year	1	Indirectly assessed as shown in the Appendix
	Total	14	

Table 2. Rating of the south African system for registering Ebola virus disease (EVD) both as a multiple cause of death and as a notifiable disease based on 2015 data.

includes EVD (ICD-10: A98.4). According to the South African disease notification system, any notifiable disease resulting in death must be doubly notified, first as a case and second as death. Thus, if one of these notifiable hemorrhagic fevers occurred, it would result in death.

It should be reflected in either the disease surveillance system or the causes of death data from the CR/VS system. If any cause of death is recorded in the CR/VS system, an analysis of the data using the multiple-cause approach has more chance of detecting the cause than the one based on underlying-cause approach.

Using the criteria developed above, the rating of the South Africa system for registering EVD both as a multiple cause of death and as a notifiable disease is given in **Table 2**. Based on latest available data at the time of writing (2015 data), The total score for South Africa is 14 of the maximum of 23. According to the definition defined earlier, this is 'fair' (in between 'poor' and 'good'). While the multiple causes of death component are excellent, the overall ranking is rated down because of the lesser performance of the surveillance system.

In **Table 2**, the dimension 3.2 could not be assessed directly as there is no case of EVD for latest year 2015. Since in the absence of outbreaks EVD is a very rare disease, one would need to collect data over several years to enable the comparison of the two systems. This is done in Appendix 1.

4. Discussion

The major challenge faced in this chapter is the irony in which the countries affected by EVD are the same ones with weak vital registration systems that are neither likely to collect data using ICD-10 nor likely to submit causes of death data to the WHO. Of all African countries, only South Africa collects and publishes data on multiple causes of death and has been doing so since 1997. But in South Africa, EVD is very rare. This rarity plus the use of three-digit ICD-10 coding in the early 2000s frustrated attempts at comparing data on EVD based on disease notification and those based on multiple causes of death. An indirect approach had to be used based on another viral hemorrhagic fever, which is endemic to South Africa, CCHF [10]. This indirect approach helped to establish complementarity of the disease surveillance system and the multiple causes of death statistics system.

The chapter has tried to argue that the system of multiple causes of death complements that of disease notification. Under the ideal conditions, for highly fatal notifiable diseases, the number of cases reported in the disease notification system should be close to the number of deaths due to that disease when reported as a multiple cause of death. This complementary relationship has several implications. The first is that, in the early stages of the development of two systems, one can be used to check on the accuracy of the other. The second is that, through record linkage methods, the data from the disease surveillance system can be linked with the data from the multiple causes of data for more in-depth analysis. The third is that the spread of the disease can be better understood through analysis of the place of disease notification against the place of death as obtained from the death statistics. The fourth is that both systems help to establish accurate endemic levels against that to gauge the start of epidemics.

With some concerted efforts, African countries can set up the ideal systems described in this chapter. Some recommendations for doing so are as follows:

1. Exploit the use of mobile phones (mHealth) in the disease notification process.
2. Embark on training the trainer program by selecting a few officers with medical background (e.g., nursing) and train them in ICD-10 coding. Through request for training assistance, one experienced trainer can be invited to come and train the trainers.
3. As automatic coding software are available free of charge, the software can be obtained, and through request for training assistance, one experienced trainer can be invited to come and train the trainers on using the software
4. Again through request for training assistance, one experienced official statistics officer can assist the African countries in analyzing data on multiple causes of death.

5. Conclusion

The chapter has shown how to set up systems capable of registering EVD both as a notifiable disease and as a multiple cause of death. The chapter has given arguments in favor of the benefits of such systems. Through a program of training the trainers, it is possible for African countries to achieve this within a few years, if concerted efforts are made.

Appendix 1

Comparing data from the disease notification system and the multiple causes of death statistics system in South Africa

To compare data from the disease notification system and the multiple causes of death statistics system, we need (1) comparable period, (2) comparable geography and (3) comparable diseases (causes of death). Since EVD is a rare disease in South Africa, a group of years should be chosen instead of a single year. For this purpose, the period 2000–2005 has been chosen. For this period, causes-of-death coding is done at the three-digit ICD-10 level. The three-digit level, as opposed to the four-digit level, loses some specificity in disease classification. For example, since the four-digit code for EVD is A98.4, under three-digit coding, this is appropriately coded as A98 ('Other viral hemorrhagic fevers, not elsewhere classified'). This category includes the following, Crimean-Congo hemorrhagic fever (CCHF) (A98.0), Omsk hemorrhagic fever (A98.1), Kyasanur hemorrhagic fever (A98.2), Marburg hemorrhagic fever (A98.3), Ebola virus disease (A98.4) and hemorrhagic fever with renal syndrome (A98.5). Of these above-mentioned hemorrhagic fevers, only CCHF is endemic to South Africa [12].

Based on the data available, the closest comparison one could make is between disease notification for CCHF (A98.0) and multiple causes of death due to 'Other viral hemorrhagic fevers, not elsewhere classified' (A98). The data used in the comparison are data from the South African disease surveillance system for 2000–2005 and the national vital registration data on causes of death for the same period.

Following [14], if $n_{ab}^{cd(j)}$ represents the number of deaths belonging to the sex *a*, age group *b*, with underlying cause *c*, a multiple cause *d* whose order of mention is *j*, then 'any mention' of a specific cause, d (irrespective of position of mention) is given as:

$$n_{\bullet\bullet}^{\bullet d(\bullet)} = \sum_{i=1}^{N}\sum_{j=1}^{m} k_i^{d(j)}$$

where *i* stands for any record out of N death records and k_i^x 's are indicator variables defined as follows:

$$k_i^{d(j)} = \begin{cases} 1 \text{ when } d = d^*(\text{the selected multiple cause (s) with order of mention } j) \\ 0 \text{ otherwise} \end{cases}$$

where j = 1,…m (the maximum number of causes per death) [14]

This expression makes up the core of the software Cause-limp 1.1 used for extracting the multiple cause data from the death records. The records were searched to any mention of 'Other viral hemorrhagic fevers, not elsewhere classified') (A98). The variables used in the analysis are the following: year of death, sex, and all the multiple causes listed on the certificate (five causes in all) including the underlying cause of death. The program routinely eliminates all still births and all those with missing recording of sex. It is restricted to those whose place of residence and death is South Africa.

Over the study period, 2000–2005, the total number of reported deaths analyzed was over 2.7 million (2,702,710). This was the number of records remaining after eliminating still births, and a number of deaths of unknown and unspecified sex were eliminated.

Of these records analyzed, the total number of deaths with any mention of hemorrhagic fever (A98) was 12. In 2000, seven deaths were recorded with A98 as a multiple cause of death. For each of the remaining years, only one death was recorded. The number of deaths with A98 as a multiple cause was highest in Free State (five), while only one multiple cause death was recorded for each of the provinces, with the exception of Limpopo where no multiple cause death was recorded.

There is a very little chance that these fevers could be of Asiatic origin (e.g., Omsk hemorrhagic fever (A98.1) or Kyasanur Forest disease (A98.2)). As mentioned in a South African manual on hemorrhagic fevers, 'Omsk hemorrhagic fever is a tick-borne virus of Siberia and Kyasanur forest disease is a tick-borne virus of the Indian subcontinent. These infections are unlikely to be seen in Africa' [15]. This makes it very likely that the disease could be CCHF as it is the only endemic one in the remaining list. As the two figures are comparable, this confirms that the two systems are comparable.

Source: Reformatted output from Cause_limp v 1.1 [16].

Over the same period, the number of notifications for CCHF was 21, and of these, the number that died was 11, close to what was obtained above based on the CR/VS system.

	Province of residence of deceased									
	Gauteng	**Free State**	**Northern Cape**	**Western Cape**	**Eastern Cape**	**Mpuma-langa**	**Limpopo**	**North West**	**KwaZulu-Natal**	**SA**
2000	1	2	1	0	1	1	0	1	0	7
2001	0	1	0	0	0	0	0	0	0	1
2002	0	1	0	0	0	0	0	0	0	1
2003	0	0	0	0	0	0	0	0	1	1
2004	0	1	0	0	0	0	0	0	0	1
2005	0	0	0	1	0	0	0	0	0	1
TOTAL	1	5	1	1	1	1	0	1	1	12
Total death records	504,425	249,430	60,972	218,908	374,387	202,084	290,924	242,230	628,559	2,702,710

Source: Reformatted output from Cause_limp v 1.1 [16].

Table A1. Trends in multiple causes of death in South Africa due to hemorrhagic fevers (ICD-10: A98) for both males and females in different provinces of residence, 2000–2005.

	Gauteng	**Free State**	**Northern Cape**	**Western Cape**	**Eastern Cape**	**Mpuma-langa**	**Limpopo**	**North West**	**KwaZulu-Natal**	**All cases**	**All deaths**
2000	1	3	2	0	1	0	0	1	0	8	5
2001	0	1	2	1	0	0	0	1	0	5	2
2002	0	1	2	0	0	0	0	0	0	3	1
2003	0	0	0	0	0	0	0	0	0	0	0
2004	0	1	1	0	0	0	0	2	0	4	2
2005	0	0	0	1	0	0	0	0	0	1	1
TOTAL	1	6	7	2	1	0	0	4	0	21	11

Source: [11].

Table A2. Trends in laboratory confirmed cases of Crimean-Congo hemorrhagic fever (CCHF) (A98.0) for both males and females in different provinces, 2000–2005.

Author details

Sulaiman Bah

Address all correspondence to: sbah@ud.edu.sa

Department of Public Health, College of Public Health, Imam Abdulrahman Bin Faisal University, Saudi Arabia

References

[1] Kennedy SB, Nisbett RA. The Ebola epidemic: A transformative moment for global health. Bulletin of the World Health Organization. 2015;**93**:2

[2] WHO Ebola Response Team. Ebola Virus Disease in West Africa–The First 9 Months and the Epidemic and Forward Projections. The New England Journal of Medicine. October, 2014;**371**:16

[3] Schieffeliin J, Shaffer J, Goba A, et al. Clinical illness and outcomes in patients with Ebola in Sierra Leone. The New England Journal of Medicine. November, 2014;**371**:22

[4] The Lancet. Ebola–A failure of international collective action. The Lancet. August, 2014;**384**

[5] Center for Disease Control. Outbreaks Chronology: Ebola Virus Disease. [Online] [Cited: September 26, 2014] http://www.cdc.gov/vhf/ebola/outbreaks/history/chronology.html

[6] Frame D, Casals J, Dennis E. Lassa virus antibodies in hospital personnel in western Liberia. Transactions of the Royal Soceity of Tropical Medicine and Hygiene. 1979;**73**:2

[7] Knoblock J, Albiez E, Schmitz H. A serological survey on viral haemorrhagic fevers in Liberia. Annales de l'Institut Pasteur/Virologie. 1982;**133**(2):125-128

[8] Khan S et al. New opportunities for field research on the pathogensis and treatment of Lassa fever. *Antiviral Research*. 2008;**78**(1):103-115

[9] Ahmadi A et al. Disease surveillance and private sector in the metropolitans: A troublesome collaboration. Int J Prev Med. 2013;**4**(9):1036-1044

[10] Swanepoel R, Struthers JK, Shepherd AJ, McGillivray GM, Nel MJ, Jupp PG. Crimean-congo hemorrhagic fever in South Africa. The American Journal of Tropical Medicine and Hygiene. 1983;**32**:6

[11] Benson FG, Musekiwac A, Blumberg L, Rispele LC. Comparing laboratory surveillance with the notifiable diseases surveillance system in South Africa. International Journal of Infectious Diseases. 2017;**59**:141-147

[12] van Vuren, Petrus J, et al. Viral haemorrhagic fever outbreaks, South Africa, 2011. The Communicable Diseases Surveillance. Bulletin. April 2012;**10**(1):1-5

[13] Stats SA. Mortality and Causes of Death in South Africa, 2015: Findings from Death Notification. Pretoria: Stats SA; 2017

[14] Bah S, Mahibbur Rahman M. Measures of multiple-cause mortality: A synthesis and a notational framework. Genus. 2009;**LXV**(2):29-43

[15] Swanepol R. Recognition and Management of Viral Haemorrhagic Fevers: A handbook and resource directory. Sandringham: National Institute for Virology;1987

[16] Bah S. Cause_Limp v1.0: A windows-based software for analyzing data on multiple causes of death. Journal of Health Informatics in Developing Countries. 2009;**3**(2):1-4

Section 2

Management of Ebola

Chapter 3

Outcomes in Baby Deliveries among Pregnant Ebola Survivors

Wen-Ta Chiu, Jonathan Wu, Stanley Toy,
Rachele Hwong, John J Stewart and Jennifer Chang

Additional information is available at the end of the chapter

http://dx.doi.org/10.5772/intechopen.74669

Abstract

Greater El Monte Community Hospital (GEMCH), the Los Angeles Department of Public Health, and the Centers for Medicare and Medicaid Services assisted in the first documented case of Ebola survivor delivery in the United States. A descriptive qualitative review of GEMCH's events and the limited documented cases of outcomes of baby deliveries among EVD survivors is discussed. Limited resources and capacity in many developing countries impact adversely on the outcomes of the EVD survivors and their neonates. Three lessons for public health workers emerge: (1) the need for the United States to strengthen their capability to manage EVD cases and other highly contagious and severe infectious diseases; (2) the revealing that EVD survivors can deliver normal, EVD free babies when using the recommended guidelines; (3) The need for health care workers to adopt and share the practical procedures in the Recommended Guidelines by the CDC and LADPH from this event are useful and can be shared with the medical fraternity. This case illustrates that EVD survivors can be equally accepted and treated with success at designated health facilities. Demystifying Ebola and eliminating social stigma surrounding the disease is crucial in this undertaking.

Keywords: Ebola virus disease (EVD), survivor, pregnancy, public health awareness, neonate, stigma

1. Introduction

The West Africa Ebola virus disease (EVD) outbreak is globally recognized as one of the largest and most severe Ebola epidemic. The disease affected over 28,000 people and the case mortality rate was estimated to be around 50%. Among pregnant women, EVD increases the possibility for spontaneous abortion and pregnancy-associated hemorrhage. In addition, neonatal mortality

rate has been found to be high with almost no chances of survival [1]. Vertical transmission from mother to the fetus is not well understood. It can occur during the acute EVD infection leading to intrauterine fetal death or as stillbirth, or neonatal death. Acutely infected women have high levels of Ebola viral nucleic acid persist in the amniotic fluid following the clearance of viremia; it is not known if this fluid is infectious. Outside of this acute period of infection, little is known about vertical transmission. Furthermore, research on pregnancy outcomes among mothers who have recovered from EVD is limited. The media too has influenced community perception of risk and enhanced stigma against EVD survivors. Public health awareness is needed to educate the individual and community on health issues related to EVD. As of June 2016, an estimated 17,232 people survived the West Africa EVD outbreak [2]. Among the survivors 5000 women are of childbearing age [3]. Some of these women will require obstetric care. The aim of this study is to examine the outcomes at baby delivery among pregnant EVD survivors. There has been contrasting results in developed countries versus developing countries. The care provided to EVD patients and survivors requires evaluation in order to support the patient throughout their illness and recovery. In women who have been infected while pregnant, the virus may be found in the placenta, amniotic fluid and fetus.

2. Methods

To assess the outcomes of EVD survivors giving birth, a case study from the AHMC Health System facility and desk research was conducted. The primary case includes the first documented EVD survivor to give birth in the United States. The facility worked in collaboration with the Centers for Disease Control and Prevention (CDC) and Los Angeles County Department of Public Health (LACDPH) to provide coordinated care. This case study provided an empirical review which explores the outcomes of EVD survivors giving birth in a real-time context. Desk review was chosen to supplement this primary research and involved collecting data from existing resources. The systematic review collected and summarized empirical evidence related to Ebola survivors giving birth. The database used for desk research included PubMed and WorldCat. PRISMA was used to improve the quality of the results through identification, screening, and eligibility criteria.

2.1. First documented EVD survivor giving birth in United States

A 29-year-old physician from West Africa contracted EVD in July 2014, after caring for an EVD patient in Nigeria. On July 29, the woman began having generalized feelings of malaise, joint and muscle pain. She self-administered antimalarial medications which were effective in treating arthralgia and myalgia [4]. On August 1, the woman developed a fever, and on August 3, she started vomiting and had diarrhea. The woman was admitted to an Ebola treatment center. She was isolated after receiving positive results of a real-time reverse transcription polymerase chain reaction (PCR).

According to the woman, she spent 13 days in an Ebola treatment center, where she was treated with oral rehydration therapy (fluid with modest amounts of sugar and salts used to correct dehydration), and acetaminophen (an antipyretic and pain reliever). In addition, a second course of antimalarial medications was also administered [4]. She was discharged from the Ebola treatment center on August 16, after testing negative for 2 separate EVD real-time reverse transcription PCR results. After her recovery, she developed some weariness, lethargy, loss of appetite, continued joint pain, and spot baldness. She did not report any sleep disturbances, headaches, or vision problems [4]. The woman's symptoms resolved 2–3 months later, and she fully recovered. Eight months prior to her EVD diagnosis, the woman had a spontaneous abortion at 10 weeks of gestation [4]. In January 2015, 22 weeks after her last negative EVD real-time reverse transcription PCR, she became pregnant again. The woman received routine prenatal care in West Africa. At her 25th week of pregnancy, a comprehensive ultrasound investigation was performed in Los Angeles County, California. The ultrasound assessment revealed standard fetus development [4].

In November 2015, Greater El Monte Community Hospital (GEMCH) treated this woman. GEMCH worked in cooperation with the Centers for Disease Control and Prevention (CDC) and the Los Angeles County Department of Public Health (LACDPH) throughout the patient's delivery process [5]. GEMCH identified staff members who were willing to assist during labor and delivery for the patient, and at 40 weeks and 1 day of gestation, labor was induced. Two doses of vaginal misoprostol, oxytocin, and an epidural anesthesia for pain management were administered to the patient. The woman successfully gave birth to a nine pound baby by normal delivery. The baby scored eight and nine on the Apgar scale at 1 minute and 5 minutes of age, respectively.. The mother had a second-degree perineal laceration, which was repaired. The mother and her baby (**Figures 1** and **2**) were discharged from the hospital 36 hours postpartum [5]. The pair was

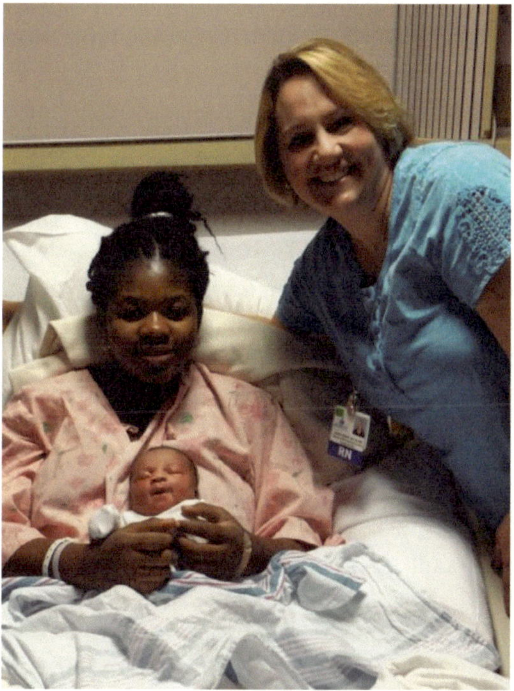

Figure 1. First documented mother EVD survivor in USA with new born baby and nurse, 2015.

Figure 2. GEMCH staff with first documented mother EVD survivor in USA with new born baby, 2015.

monitored for 6 weeks following the delivery, after which they returned home to West Africa. This case should raise awareness that EVD survivors can deliver healthy normal babies free of EVD.

Upon delivery, the Centers for Disease Control and Prevention (CDC) and Los Angeles County Department of Public Health (LACDPH) examined the mother's blood and the baby umbilical-cord blood. Oral and vaginal swabs were also sampled. Additional specimens were obtained from breast milk, first meconium, the placenta, and the amniotic fluid. All samples tested negative for EVD [5]. One week prior to the delivery, EVD real-time reverse transcription PCR testing was carried out on the patient's blood by both the LACDPH's laboratory and the CDC's Viral Special Pathogens Branch [5]. Both tests were reported negative for EVD. However, Ebola serum antibodies were detected.

2.2. Liberia EVD survivor

In 2014, a 26-year-old Liberian nurse assistant was recognized by Time Magazine. She was and awarded *Persons of the Year 2014* by *Time Magazine* in recognition of her efforts to fight EVD. The woman had contracted EVD in the summer of 2014 and experienced the acute phase of EVD alongside her sister and fiancé, now husband, at the time [6]. She was discharged with Ebola free status from the Ebola Treatment Centre on August 28, 2014. Médecins Sans Frontières hire her as mental health counselors in their Ebola units [7]. The woman became pregnant with her fourth child in the summer of 2016, 2 years after her initial contraction of EVD. She experienced a difficult pregnancy which included gestational hypertension. In February 2017 she delivered by a cesarean section at Eternal Love Winning Africa Hospital (ELWA). She was discharged a few hours later. Soon after arriving home, the woman collapsed and began frothing at her mouth. Her spouse swiftly returned her to ELWA Hospital on the evening of February 19, 2017. She developed convulsions and seizures before admission [7]. With some hesitation she was eventually admitted to the hospital. She died on February 21, 2017, just 4 days after giving birth. She is survived by her husband and four children [8]. It is not known if her fourth child has been tested for the EVD antibodies.

2.3. Sierra Leone EVD survivor

A similar pregnancy was reported in Sierra Leone in ac EVD survivor. She contracted the EVD infection when receiving gynecological services for her first pregnancy from a nurse, who had some physical contact with an infected person in Guinea during a memorial service. The first trimester of her gestation had complications which led into a miscarriage in May 2014. She was subsequently diagnosed with EVD [3]. She was discharged from the hospital on June 8, 2014 as Sierra Leone's first EVD survivor. Months after her recovery, she got pregnant again. She gave birth to a healthy baby boy on August 9, 2015. The baby was fed on formula milk substitutes. Only weeks after the baby was born, he developed a fever and died of an infection apparently not related to EVD. .

2.4. Other reported baby deliveries among Ebola survivors

Only one woman in the acute phase of EVD is reported to have given birth to a live, full term baby. This mother developed fever 4 days prior to delivery and the baby also developed fever and died 3 days later [9]. In the EVD outbreak at Mission Hospital in Yambuku, it was reported that 11 live babies were delivered from mothers in the acute stage of their EVD infection [9]. All babies had died in what appears to be neonatal mortality related to EVD. No other pathological data was available. Seven out of 10 babies were reported to have developed fevers. The route of infection from mother to baby is not clear. It is likely transmission occurred during gestation, birth, or through breast milk.

Although the virus has been isolated in breast milk, the risk of transmission must be balanced against the risk of malnutrition and other infections. If there is a safe alternative breastfeeding is not advisable [9, 10]. Neonates breastfed by mothers in the acute phase of their EVD infection have been reported to have become infected with EVD. Lactating women who are sick with EVD should be placed into isolation until recovery. Safer feeding options such as safe pasteurized donor milk or Ready to Use Infant Formula (RUIF), should be given to infant when available [3]. Breastfeeding shortly before convalesce of EVD may still expose the newborn to unnecessary risk. If available, laboratory testing of breast milk should be done to evaluate the EVD content of the milk, and if found negative breastfeeding can commence. Breastfeeding after full recovery from EVD is relatively safe and should be encouraged as there are many benefits of breast milk.

3. Discussion

3.1. Addressing stigma

There is always social stigma surrounding individuals infected with EVD and regnant Ebola survivors are certainly at risk [11]. In major outbreaks thinly veiled prejudice is frequent [12]. The isolation of patients and the associated ostracism impacts the quality of life for those affected and their communities. Furthermore, the poor level of awareness about transmission of the infection further undermined confidence and trust [11]. Mass communication and social media influenced community perception during the 2014 EVD epidemic [13]. The potential international hazard posed by EVD generated extensive media attention [3]. Structural deficiencies such as poverty, lack of education, and political conflicts further undermined the response and fueled stigma around EVD [14]. These issues fused with cultural norms led to mistaken beliefs and behavior regarding EVD and its mode of transmission [14]. Misguided understanding and stigma in individual communities also led to harassment, rejection and persecution [11].

Health providers need to support patients and those affected to access care and management free of stigma and discrimination. Emergency preparedness is needed to promote an enabling environment to overcome misconceptions and overcome related fear. Healthcare workers need to recognize the fear highly contagious diseases create, rather than treat those infected or are recovering from the disease as pariahs, need to overcome that apprehension to create preparedness. This case illustrates the need for health care providers to advocate for Assurance of the basic rights of the patients irrespective of the medical condition is essential in managing emerging infections like Ebola with high mortality. Ebola survivors should equally be accepted and managed at health facilities.

3.2. Promoting public education

Health education is critical in reducing misconceptions and fear. Initially, no hospital was willing to volunteer to admit this mother despite an appeal from the CDC and LACDPH. This would be the first documented EVD survivor to deliver in the USA. However, through partnership with CDC and LACDPH, the staff at GEMCH was educated about transmission and infection control and personal protection prior to delivery. With proper knowledge, public education and reassurance, it was possible to mobilize interdisciplinary teams for patient management. The United States is strengthening their capacity to manage EVD cases and other highly contagious and severe infectious diseases. In February 2015, 55 hospitals were designated as Ebola treatment centers. In June 2015, the Department of Health & Human Services (HHS) adopted CDC guidelines and identified and funded nine health facilities as regional EVD treatment centers for specialized emergency medical care. In California, for instance, Cedars-Sinai Medical Center serves as the regional treatment center and will serve Arizona, Hawaii, Nevada, and the U.S. territories in the Pacific.

3.3. Enhancing continuing medical training

A study published in the American Journal of Infection Control analyzed conventional training programs and reinforced training programs with their ability to decrease self-contamination during the usage of basic PPE and enhanced PPE. The reinforced training programs provided evidence of improvement in adhering to protocol and proficiency. However, there is still apprehension on a perfect barrier to infection contamination as this remains elusive [15]. According to the findings from the Ebola cases introduced, it is crucial for healthcare providers to work in partnership in educating health workers and patients. The treatment must be provided in a manner that promotes patient dignity and comfort at the appropriate level of care. Healthcare facilities need to bring awareness to their physicians and medical staff in regard to pregnant Ebola survivors who have fully recovered. These patients pose virtually no risk to others including the neonates [16]. Thus healthcare workers can safely treat EVD survivors.

Recommended Guidelines.

Below is a summary of the Recommended Guidelines provided by the CDC and LADPH directly drawn from the GEMCH Ebola survivor delivery event:

a. *Before delivery*: For pregnant patients with intact membranes, medical staff is not required to utilize Personal Protective Equipment (PPE). Routine hand hygiene before and after contact with the patient are required.

b. *For vaginal exams with rupture membrane*: face protection with a face mask and face shield, fluid-resistant gowns, and gloves are required to prevent mucous membranes and splashes exposure.

c. *For obstetrical and additional procedures where a large amount of fluid or blood is anticipated*: such as artificial rupture of membranes and postpartum hemorrhage, PPE is required.

d. *During the delivery process*: medical staff must wear PPE for the patient and staff's safety. Prevent exposure from mucous membranes, body fluids, and splashes, utilize face protection with a face mask and face shield, fluid-resistant gowns, double gloves, and boot covers that extend to at least mid-calf, must be utilized. Pockets and surgical drapes should be placed to prevent dousing of gowns and boots from body fluids during the delivery process.

e. *After the delivery:* medical staff should not remove PPE until the mother's gown and bedding have been changed.

f. *Post delivery:* Upon vaginal and perineal examinations, non-sterile exam gloves and disposable gowns must be utilized. Face protection is not needed unless the occurrence of body fluid splash.

g. *During initial contact with the baby*: medical staff must apply PPE when giving them a bath. If exposure to fluids when changing diaper and removing waste, non-sterile exam gloves and two face masks must be worn. At all cost, skin-to-skin contact should be delayed until baby is bathed.

There has been contrasting results in There is some contrast in outcomes of EVD survivors and the neonate between developed and developing. The treatment offered to survivors who become pregnant must be assessed and adapted to the different resource environments and should be consistent to support the patient throughout the delivery process. Positive outcomes can be achieved through the application of recommended guidelines so that EVD survivors as well as the neonates may be assured of their future.

4. Conclusion

The Ebola global threat has receded at least for now, but the world must remain vigilant. The challenges created should also provide opportunities for building capacity for early detection and control. Health systems for surveillance and human resource development must be reinforced as EVD and other highly infectious diseases may re-emerge in low resource settings without adequate capacity for timely containment. A global response is required to improve health care systems in all the affected regions [17]. The care provided to EVD patients must be evaluated to support the patient throughout their illness. If the mortality rate can be lowered through the application of the standard of care guidelines, EVD patients and survivors will cooperate in l more seeking care. A number of medical problems have been reported in survivors, including mental health. Ebola survivors need comprehensive social support for the medical, mental and psychosocial challenges they face. In developing countries resources are limited. For such instances early detection and action is vital to Ebola containment. Further research is required to monitor the Ebola outbreaks as they occur in order to gain true insight to developing delivery management needs of EVD survivors especially during baby delivery.

Acknowledgements

The authors would like to thank all health facilities and staff involved with the research and writing of this manuscript. We would personally like to thank Adaora Igonoh for allowing us to share her story and bring to light EVD survivors can deliver normal, EVD free babies with standard growth and development. The authors would like to thank the staff at GEMCH including but not limited to Amanda Kamali, Moon Kim, Denise Jorison, Pierre Rollin, Okoi Lilian, Julius Kpaduwa, Evelyn Calubaquib, Jose Ortega, Stella Kpadowa, Yang Cho, Ifeoma Uisoha, and Sarah Schrier. Special appreciation to the Centers for Disease Control and Prevention (CDC) and Los Angeles County Department of Public Health (LACDPH) for their help and collaboration in this unique case. Furthermore, the authors would like to thank our system's A Team: Linda Marsh, Iris Lai, Patrick Petre, Philip Cohen, Karen Price-Gharzeddine, Rick Castro. We would also like to thank the corporate project team members Su-Yen Wu, Claire Huang, Molly Wu, Qian (Eowyn) Wen, Sasha Yu, Ping Lin.

Author details

Wen-Ta Chiu[1,2], Jonathan Wu[1], Stanley Toy[1], Rachele Hwong[1]*, John J Stewart[1] and Jennifer Chang[1]

*Address all correspondence to: rachele.hwong@ahmchealth.com

1 AHMC Health System, Alhambra, United States

2 Taipei Medical University, Taipei, Taiwan

References

[1] Jamieson D. What obstetrician-gynecologist should know about Ebola: A perspective from the Centers for Disease Control and Prevention. Obstetrics and Gynecology. 2014; **124**(5):1005-1010. DOI: 10.1097/AOG.0000000000000533

[2] Situation Report. World Health Organization. 2016. Retrieved September 29, 2017, from http://apps.who.int/iris/bitstream/10665/208883/1/ebolasitrep_10Jun2016_eng.pdf?ua=1

[3] Ebola (Ebola Virus Disease). 2016, June 10. Retrieved September 29, 2017, from https://www.cdc.gov/vhf/ebola/outbreaks/2014-west-africa/survivors.html

[4] Kamali A, Jamieson DJ, Kpaduwa J, Schrier S, Kim M, Green NM, Mascola L. Pregnancy, labor, and delivery after Ebola virus disease and implications for infection control in obstetric services, United States. Emerging Infectious Diseases. 2016;**22**(7):1156-1161. DOI: 10.3201/eid2207.160269

[5] Chiu WT, Toy S, Wu J. The first Ebola survivor to deliver a full term baby in the United States. American Public Health Association. 2016;**106**(6):e4-e5

[6] McVeigh K. Ebola Survivor and Frontline Fighter Dies after Childbirth Complications. 2017, February 28. Retrieved September 29, 2017, from https://www.theguardian.com/global-development/2017/feb/28/ebola-survivor-and-frontline-fighter-dies-after-childbirth-complications

[7] Krausz J. Salome Karwah Dies: Liberian Ebola Fighter Loses Life After Childbirth. 2017, February 27. Retrieved September 29, 2017, from http://www.newsmax.com/TheWire/salome-karwah-dies-ebola-fighter/2017/02/27/id/775880/

[8] Why Ebola Fighters Are TIME's Person of the Year 2014. n.d. Retrieved September 29, 2017, from http://time.com/time-person-of-the-year-ebola-fighters-choice/

[9] Richens Y, Gerrard J. Ebola in pregnancy. British Journal of Midwifery. 2014;**22**(11): 771-775. DOI: 10.12968/bjom.2014.22.11.771

[10] Fischer WA, Hynes NA, Perl TM. Protecting health care workers from Ebola: Personal protective equipment is critical but is not enough. Annals of Internal Medicine. 2014;**161**: 753-754. DOI: 10.7326/M14-1953

[11] Davtyan M, Brown B, Folayan MO. Addressing Ebola-related stigma: Lessons learned from HIV/AIDS. Global Health Action. 2014. DOI: 10.3402/gha.v7.26058

[12] Sharfstein JM. On fear, distrust, and ebola. JAMA. 2015;**313**(8):784-784. DOI: 10.1001/jama.2015.346

[13] Nagpal SJ, Karimianpour A, Mukhija D, Mohan D. Dissemination of 'misleading' information on social media during the 2014 Ebola epidemic: An area of concern. Travel Medicine and Infectious Disease. 2015;**13**(4):338-339. DOI: 10.1016/j.tmaid.2015.05.002

[14] Gates B. The next epidemic – lessons from Ebola. The New England Journal of Medicine. 2015;**372**(15):1381-1384. DOI: 10.1056/NEJMp1502918

[15] Casalino E, Astocondor E, Carlos J, Diaz-Santana DE, Aguila CD, Carrillo JP. Personal protective equipment for the Ebola virus disease: A comparison of 2 training programs. American Journal of Infection Control. 2015;**43**(12):1281-1287

[16] Ebola Virus Disease in Pregnancy: Screening and Management of Ebola Cases, Contacts and Survivors. 2015. World Health Organization

[17] Baden LR, Kanapathipillai R, Campion EW, Morrissey S, Rubin EJ, Drazen JM. Ebola–an ongoing crisis. The New England Journal of Medicine. 2014;**371**(15):1458-1459. DOI: 10.1056/NEJMe1411378

Section 3

Drugs and Vaccines

Chapter 4

Predicting Candidate Epitopes on Ebola Virus for Possible Vaccine Development

Maryam Hemmati, Ehsan Raoufi and Hossein Fallahi

Additional information is available at the end of the chapter

http://dx.doi.org/10.5772/intechopen.72413

Abstract

Zaire ebolavirus, a member of family *Filoviridae* is the cause of hemorrhagic fever. Due to lack of appropriate antiviral or vaccine, this disease is very lethal. In this study, we tried to find epitopes for superficial glycoprotein and nucleoprotein of *Zaire ebolavirus* (that have high antigenicity for MHC I, II and B cells) by using *in silico* methods and immunoinformatics approach. By using CTLPred, SYFPEITHI and ProPred web applications for MHC class I and SYFPEITHI and ProPred1 web applications for MHC class II, we had been able to find epitopes (peptides) that have the highest score. Also ElliPro, IgPred and DiscoTope web tools had been performed to predict B cells conformational epitopes. Linear epitope prediction for B cell was performed with six methods from IEDB. All of the results that including candidate epitopes for T cells and B cells were reported. It was expected that these peptides could be stimulated immune response and used for designing the multipeptide vaccine against ZEV but these results should be reliable with experimental analysis.

Keywords: epitopes, glycoprotein, immune response, immunoinformatics approach, multipeptide vaccine, nucleoprotein, *Zaire ebolavirus*

1. Introduction

Zaire ebolavirus, a member of genus *Ebola virus,* family *Filoviridae,* and order *Mononegavirales,* is enveloped, RNA negative strand genome and filamentous virus. This virus is the cause of serious hemorrhagic fever (HF) in human and the mortality rate is 50–90% [1, 2]. There have been several outbreaks of *Ebola virus* up to now and the last one was on 2014.This outbreak started in Guinea and spread to other countries in West Africa. The 2014 outbreak was the largest one which had the highest mortality rate and risk of spreading to different parts of the world [3, 4].

Experimental work with EBOV is very difficult and currently, there is no effective and licensed vaccine available for Ebola. Vaccination is a good approach for the prevention and treatment against many types of diseases like the viral infection. There are several types of vaccines, and the goal of all of them is presenting antigenic fragment to the immune system to induce the adaptive immune response. Several types of antiviral vaccines consist of inactivated, live attenuated, virus-like particles and DNA vaccines. Another type of vaccine is peptide-based vaccines. These vaccines are based on designing different epitopes for B cells and T cells. These vaccines, in comparison with other vaccines, have fewer side effects and are safe and easy to prepare [4–6]. Obviously, T cells have an important role in stimulating immune responses, but for having a better and accurate response, first of all, antigenic fragments should be attached to major histocompatibility complex (MHC) molecules. MHC molecules process and present antigenic peptides to T cells. These peptide epitopes must be linear for attaching to MHC molecules [7]. The most of T cells are belonging to two groups namely CD8+ and CD4+. The difference between two groups refers to different glycoproteins in the surface of T cells. CD8+ T cells are cytotoxic T cells (CTL) that bind to MHC class I molecules and CD4+ T cells are T helper cells that attach to MHC class II [8, 9]. B cells are another part of immune systems, which have receptors and secrete antibodies. Epitopes can be discontinuous or continuous for B cells. These epitopes can also bind to lipids, carbohydrates and peptides but 3D structure of antigens has an important role in stimulating humoral immune response by B cells [10]. With the advances in in silico method and bioinformatics, immunoinformatics, a branch of science was progressed. Immunoinformatics is an interdisciplinary science that emanates from immunology and bioinformatics and generates meaningful immunological data. With the help of immunoinformatics approaches, we can find epitopes for B cells and T cells, and also we can design vaccines based on peptide or multipeptide vaccine [4, 11].

In this chapter, we tried to use immunoinformatics tools and in silico method to predict MHCs linear epitopes and B cells discontinuous and linear epitopes (peptides) for glycoprotein and nucleoprotein of *Zaire ebolavirus* (as antigen). These results can be useful for finding epitopes that can be the candidate for designing vaccine and therapeutic strategies for fighting with HF (hemorrhagic fever) of Ebola virus.

2. The best candidate for producing vaccines against *Ebola virus*

The ZEOBV genome has seven ORF, including NP-VP35-VP40-GP-VP30-VP24-L. Nucleoprotein (NP) causes encapsulation of the genome of Ebola. It has been aggregated with VP30 (transcription factor) and VP35 (polymerase cofactor). L is RNA-dependent RNA polymerase. VP24 is a minor matrix protein that associates with the membrane. VP40 is a major matrix protein that can mediate virus particles creation [1, 4, 12]. Glycoprotein (GP) is present on the surface of Ebola. Secreted nonstructural GP, structural GP 1 and GP 2 are the results of mRNA editing during transcription of GP of ZEBOV. The N-terminal sequence between the GP and sGP is the same but the C-terminal of them is different [4, 13]. The ratio between sGP and GP during infection is 80–20%. Superficial GP is cleaved by furin in Golgi to GP1 and GP2 that form homotrimeric proteins on the surface of EBOV

[5, 10]. The GP1 is a soluble protein on the surface of virus, which has a mucin-like domain and is highly glycosylated.GP2 is a membrane-spanning subunit and smaller than GP1, which is connected to GP1 by disulfide bonds [14]. GP has a major role in attaching to host cells and has cytotoxicity effects. This polycistronic GP is the main difference between ZEBOV and other mononegavirus. Because of the GP on the surface, this protein has the most antigenicity for the immune system and so it can be the best candidate for producing vaccines against EBOV [2, 4]. Immunoinformatic studies on Ebola virus proteins also have shown that there is a high rate of immune response for epitopes of Ebola nucleoprotein (NP), in which, it can be even more than the response for epitopes of Ebola glycoprotein [15–18]. Therefore, it can be another target for peptide-based vaccine designing against Ebola virus.

Bioinformatics evaluation on the superficial glycoprotein showed less than 30% identity with family and less than 50% with species of identity in the this protein, this glycoprotein is also glycosylated. Studies have been shown that the structure and sequencing of GP are constantly changing and these changes are the major causes of the weakness of the immune system against the virus and, consequently, its pathogenicity. For these reasons, designing and development of the vaccine against the Ebola virus is difficult and complex.

3. Epitope mapping for the purpose of peptide-based vaccine design

The most commonly used assimilation and the concept of the term epitope mapping is the mapping of the antigenic regions detected by antibodies. This term is ambiguous because the purpose of using this term in research is not clear, and this mapping can be done with a variety of goals. Therefore, researchers have tried to use alternative terms for epitope mapping in articles in order to clarify the purpose of this bioinformatics process such as immunological analysis, microstructure analysis and epitope determination. Despite all these efforts and including a wide range of goals, scientists are still advised to use this term in articles and resources.

A number of important applications of epitope mapping including:

- Determination of biological process mechanism,
- Recognition of an epitope of practical value,
- Connection of any type of polymorphism such as SNP to protein structure or Ab binding,
- Characterization of Ab binding in patients,
- Evaluation of vaccine design,
- Identification of autoimmune diseases,
- Qualification of an Ab for diagnostic use such as Western blot analysis, trans-species assays, finding isoforms of Ag and allergen characterization,
- Distinguishing of antigen peptide mimic,

- Establishing of antigen structure and
- Finding an antigen which it can make different between immunization antibodies and infection antibodies.

Immunogenicity studies have shown epitope mapping (the branch of bioinformatics) is based on mathematical modeling and evolutionary algorithms. Epitope mapping methods can be used to molecular modeling behaviors in nature, so the data obtained from these methods can predict molecular interactions, such as binding of antigen to antibody and peptide to MHC, with high probability [19, 20].

Designing of vaccines based on peptides or epitopes is the goal that we have been considering in this study. They are within the epitope mapping compass and to achieve these goals, we identify and predict the epitopic areas by using approved and high-performance software.

Mathematically modeled methods for epitope mapping are artificial neural networks (ANN), quantization matrices, decision trees, HMM (Markov secret models), SMM is a stabilized matrix method (SMM), among which ANN, SVM and HMM are capable of analyzing linear and nonlinear data. In many applications that analyze epitope prediction, these three methods are used to identify the sequential linear epitopes of the T lymphocytes and the nonlinear (spatial) epitopes of the B lymphocytes [4, 21, 22].

There are other predictive methods for detecting spatial epitopes of the B cell lymphocytes, which include the following: homology modeling, docking, 3D- and threading techniques. These methods with the ANN, SMM and HMM methods are most used in the study of the spatial epitope and are among the software main methods used in this research [4, 23]. In sum, all methods for identifying epitopes that have the best antigenic properties but the epitopes should have important characteristics, including (1) structural flexibility, (2) in the surface of the protein, (3) exposure to solvent, (4) containing charged amino acids and (5) contains hydrophilic amino acids.

4. Candidate for designing peptide-based vaccines against *Ebolavirus*

Conventional vaccines are prepared from an attenuated or inactive version of the pathogen. However, often the antigen to which the immune system responds is of small number of amino acids or peptide of antigen. The alternative approach to stimulate the immune response, such as humoral and cell-mediated immune response, is the identification of peptide sequences or epitopes that have a protective immune response. Therefore, these epitopes can have no risk of mutation and they would be more stable, and also, there can be fewer side effects [24, 25]. This approach is called peptide-based vaccine designing.

4.1. Determine the T cell antigenic fragment

Epitope analysis was performed for MHC classes I and II.

http://dx.doi.org/10.5772/intechopen.72413

4.1.1. Epitopes analysis for MHC class I

By using CTLPred (http://www.imtech.res.in/raghava/ctlpred/index.html), SYFPEITHI (http://www.syfpeithi.de/bin/MHCServer.dll/EpitopePrediction.htm) and ProPred1 (http://www.imtech.res.in/raghava/propred/) web tools, we had been able to find some peptidic epitopes for MHC class I.

A. Finding epitopes for MHC class I by using CTLPred web tool and sequence of GP and NP of ZEBOV. In **Tables 1** and **2**, three peptides that have higher score are shown for each protein. Positions 19, 245 and 130 for GP and positions 292, 263 [4] and 397 for NP have higher scores, so these peptides have been suggested by the application to be epitopes.

B. Using SYFPEITHI web tool for predicting epitopes. We investigated some alleles of the MHCI for the protein of interest. According to these results, as is shown in **Table 3**, for GP 564, 246 and 205 positions have a higher score in this prediction web tools [4]. But when we investigated other alleles, we had observed that the position 246 was the most repetitive of all and had a high score. For NP, results are shown in **Table 4**, position 266 has the highest score.

C. ProPred1 was used to predict epitopes of GP and NP, which bind with the highest score to MHC class I alleles. The peptide with highest score was selected and is shown in

Strat position	Sequence	Score(ANN\SVM)
130	RGFPRCRYV	0.97/0.99110928
245	ESRFTPQFL	0.98/0.867525
19	FFLWVIILF	0.97/078863896

Table 1. Epitope prediction of GP based on CTLPred web server.

Strat position	Sequence	Score(ANN\SVM)
292	EYAPFARLL	0.96/1.5915619
263	RLHPLARTA	0.89/1.2678105
397	LRKERRLAKL	0.84/1.0652382

Table 2. Epitope prediction of NP based on CTLPred web server.

Allele	Sequence	Position	score
HLA-A*26	ETTQALQLF	564	31
HLA-A*26	GVIIAVIAL	660	28
HLA-A*01	ATEDPSSGY	205	30
HLA-A*01	LFEVDNLTY	233	27
HLA-B*08	TRKIRSEEL	298	29
HLA-B*1402	SRFTPQFLL	246	30
HLA-B*1402	NRKAIDFLL	586	27
HLA-B*1516	YFGPAAEGI	534	27
HLA-B*2705	LRTFSILNR	579	27
HLA-B*37	RDRFKRTSF	11	27
HLA-B*40:01	NETTQALQL	563	27
HLA-B*58:02	RATTELRTF	574	29
HLA-A*02:01	GLICGLRQL	553	29
HLA-A*03	TVIYRGTTF	168	27
HLA-A*03	FLILPQAKK	183	27
HLA-A*03	QIIHDFVDK	625	27
HLA-A*11:01	STHNTPVYK	387	28

Table 3. Epitopes with scores above 27 of GP for MHC class I according to SYFPEITHI web server.

Tables 5 and **6**. According to this result, the highest score belonged to position 264 for the HLA-B*2705 allele for GP, and as shown in **Table 6** for NP, positions 273 and 109 for HLA-A20 allele have higher scores.

Finally with revision of all data that are achieved for MHC class I, we are able to conclude that sequences: "TRKIRSEEL" (with position 298 for T residue), "SRFTPQFLL" (with position 246 for S residue) and "ETTQALQLF" (with position 561 for E residue) for GP [4], and sequences: "RLHPLARTA" (with position 263 for first residue), "SRELDHLGL" (with position 360 for

Allele	Sequence	Position	score
HLA-A*01	TSDGKEYTY	680	29
HLA-A*01	YPDSLEEEY	688	28
HLA-A*03	PLARTAKVK	266	31
HLA-A*03	PVYRDHSEK	609	30
HLA-A*03	TVLDHILQK	249	29
HLA-A*03	EVKKRDGVK	107	28
HLA-A*03	TLRKERLAK	396	28
HLA-A*03	TVAPPAPVY	603	27
HLA-A*1101	TVLDHILQK	249	28
HLA-A*26	EVNSFKAAL	276	30
HLA-B*08	LRKERRLAKL	397	30
HLA-B*08	NHKNKFMAI	726	27
HLA-B*1402	ARFSGLLIV	239	30
HLA-B*1402	SRELDHLGL	360	29
HLA-B*4001	NEENRFVTL	707	27
HLA-B*0201	RLEELLPAV	116	28
HLA-B*0201	GLFPQLSAI	311	27

Table 4. Epitopes with scores above 27 of NP for MHC class I according to SYFPEITHI web server.

first residue) and "VKNEVNSFK" (with position 273 for first residue), have the highest score and the most frequent within these analysis. Therefore, it can be a better candidate peptide epitopes than other sequences.

4.1.2. Epitopes analysis for MHC class II

By using SYFPEITHI and ProPred web tools, some peptidic epitopes for MHC class II were predicted.

Allele	Sequence	Position	Real score
HLA-B*2705	TRKIRSEEL	298	10000
HLA-B*2705	GREAAVSHL	357	9000
HLA-B*2705	FQRTFSIPL	27	6000

Table 5. Epitope prediction from GP for MHC class I by ProPred1 web tools (only top score peptides are shown).

Allele	Sequence	Position	score
HLA-A20 cattle	VKNEVNSFK	273	4000
HLA-A20 cattle	KKRDGVKRL	109	4000
HLA-B2705	SRELDHLGL	360	3000

Table 6. NP epitope prediction of MHC class I according to ProPred1 web server.

Allele	Sequence	Position	score
HLA_DRB1*0101	CRYVHKVSGTGPCAG	135	35
HLA_DRB1*0101	FFLYDRLASTVIYRG	159	33
HLA_DRB1*0701	IILFQRTFSIPLGVI	24	34
HLA_DRB1*0101	LRTFSILNRKAIDFL	579	30

Table 7. GP epitope prediction for MHC class II based on SYFPEITHI web server.

A. Using SYFPEITHI web tools, epitope mapping of the amino acid sequence of the GP and NP of ZEBOV was performed. Accordingly, we investigated some alleles of MHC class II for the protein of interest. Peptides that have higher score were selected and are shown in **Table 7** for GP and in **Table 8** for NP. According to this result for GP, the highest score belongs to position 135 with the score of 35 and after that, the position of 159 has a higher

Allele	Sequence	Position	score
DRB1*0301	NRFVTLDGQQFYWPV	710	38
DRB1*0101	SGAVKYLEGHGFRFE	93	36
DRB1*0101	ENRFVTLDGQQFYWP	709	35
DRB1*0101	DMDYHKILTAGLSVQ	18	34
DRB1*0101	HGLFPQLSAIALGVA	310	34
DRB1*0101	MVIFRLMRTNFLIKF	198	33
DRB1*0101	HQGMHMVAGHDANDA	216	31
DRB1*0101	RHILRSQGPFDAVLY	653	31
DRB1*0701	DMDYHKILTAGLSVQ	18	32
DRB1*0107	GVDFQESADSFLLML	63	30
DRB1*1501	AGQFLSFASLFLPKL	147	32
DRB1*1501	GHMMVIFRLMRTNFL	195	30
DRB1*1501	TNFLIKFLLIHQGMH	206	30
DRB1*1101	EEMYRHILRSQGPFD	649	30
DRB1*1101	AVLYYHMMKDEPVVF	664	30

Table 8. NP epitope prediction for MHC class II based on SYFPEITHI web server.

score for the HLA-DRB1*0101 allele [4]. For NP, position 710 with score 38 have the highest score.

B. ProPred web tool was used to predict epitopes. Peptide with the highest score was selected and is shown in **Table 9** for GP and in **Table 10** for NP. This result for GP has

Allele	Sequence	Position	score
DRB1_0703	FQRTFSIPL	26	9.7000
DRB1_0817	FFLYDRLAS	158	7.2000
DRB1_0701	FQRTFSIPL	26	9.7000

Table 9. Epitope prediction of GP for MHC class II based on ProPred web server.

Allele	Sequence	Position	score
DRB1_0701	FLSFASLFL	149	7.6000
DRB1_0701	FRLMRTNFL	200	7.5200
DRB1_0701	LNLSGVNNL	299	7.2000
DRB1_0703	FLSFASLFL	149	7.6000
DRB1_0703	FRLMRTNFL	200	7.5200
DRB1_0703	LNLSGVNNL	299	7.2000
DRB1_0817	FLIKFLLIH	207	7.8000
DRB1_1501	MVIFRLMRT	197	7.7000
DRB1_1506	MVIFRLMRT	197	7.7000
DRB1_0405	FRLMRTNFL	200	7.6000

Table 10. MHC class II epitope prediction for NP of according to ProPred web server.

been shown that the positions 158 and 26 have the highest score. These positions are very frequent in reviewing other results. For NP positions, 149 and 200 have higher score according to ProPred web server.

As a result for MHC class II for GP, sequences: "IILFQRTFSIPLGVI" (with position 24 for first residue), "CRYVHKVSGTGPCAG" (with position 135 for the first residue) and "FFLYDTLAS" (with position 158 for first the residue) [4] and for NP sequences: "FLSFASLFL" (with position 149 for the first residue), "NRFVTLDGQQFYWPV" (with position 710 for first residue) and "FRLMRTNFL" (with position 200 for first residue) have the highest scores and the most frequent within these analysis.

By considering these results for MHC classes I and II, we think these epitopes can activate the cell-mediated immune response; therefore they can be used for producing peptide-based vaccines.

4.2. Determine the B cell antigenic fragment

4.2.1. Prediction of linear (sequential) epitopes

In this section, six methods from IEDB were used for the prediction of linear epitopes. A collection of methods to predict linear B cell epitopes based on sequence characteristics of the antigen using amino acid scales and HMMs. These methods include the following:

1. BepiPred Linear Epitope Prediction: BepiPred predicts the location of linear B cell epitopes using a combination of a HMM and a propensity scale method [26].

2. Chou & Fasman Beta-Turn Prediction: This method is commonly used to predict beta turns to the prediction of antibody epitopes.

3. Emini Surface Accessibility Prediction: The computation was based on surface accessibility scale on antibodies. The accessibility profile was achieved using the formulae Sn = (n + 4 + i) (0.37) − 6, where Sn is the surface probability, dn is the fractional surface probability value, and i vary from 1 to 6. A hexapeptide sequence with Sn greater than 1.0 indicates an increased probability of existing on the surface [27].

4. Karplus & Schulz Flexibility Prediction: In this method, flexibility scale based on ability to move protein segments. The calculation based on a flexibility scale is similar to classical calculation, except that the center is the first amino acid of the six amino acids window length, and there are three scales for describing flexibility instead of a single one [28].

5. Kolaskar & Tongaonkar Antigenicity: A semiempirical method that makes use of physicochemical properties of amino acid residues and their frequencies of occurrence in experimentally known segmental epitopes was developed to predict antigenic peptide established on proteins. This method can predict antigenic peptide with about 75% accuracy [29].

6. Parker Hydrophilicity Prediction: In this method, hydrophilic scale based on peptide retention times during high-performance liquid chromatography (HPLC) on a reversed-phase column was formulated [30].

4.2.1.1. Linear epitopes prediction for GP

We performed all of six methods for the GP protein sequence and summarized its results in **Table 11**.

Method number	Thereshold	Max score	The best candidate peptide sequences
1	0.271	2.51 for 384 position	370 to 394
2	1.007	1.437 for 384 postion	381 to 387
3	1	7.532 for 499 position	Scattered areas
4	1.008	1.119 for 321 position	318 to 324
5	1.015	1.225 for 183 position	180 to 186
6	1.793	7.957 for 640 position	637 to 643

Table 11. Conclusion results for B cell linear epitope prediction from GP protein sequence.

4.2.1.2. Linear epitopes prediction for NP

We performed all of six methods for the NP protein sequence and summarized its results in **Table 12**.

4.2.2. Prediction of discontinuous (conformational) epitopes

4.2.2.1. Discontinuous epitopes prediction for GP

For B cell epitope prediction, 3D structure of antigen is more important than linear sequence, therefore, in this study, we used PDB ID of GP of ZEBOV.

By considering the structure of GP in PDB, we comprehended that "I" chain of protein has a maximum length than other chains. Also, with investigating two epitopes that predicted for MHCs molecules, we understood both of them are on the "I" chain. Therefore, this chain was selected in this analysis.

A. Using ElliPro web tool, epitope mapping of the amino acid sequence of the GP of ZEBOV was performed. PDB Id (3csy) and "I" chain was used for ElliPro tool. Results from this prediction for linear epitopes have been shown that the sequence of the peptides from 31 to 64 and 255 to 310 had a higher score than the other part of the protein and are illustrated in **Figures 1–3** and **Table 13** [4].

B. By using DiscoTope and 3D structure of GP of ZEV, we could find discontinuous epitopes. These sequences may be near to each other in 3D conformation but far from each other in the amino acid sequence or the first structure. In this study, we analyze the "I" chain from GP1 and we set the threshold on −7.7, upstream regions of the threshold have the positive prediction. Two regions that are shown in **Figure 4** have positive predictions but 261–310 regions have a more score and these scores for each amino acid are illustrated in **Table 14** [4].

In **Table 14**, contact number indicates the number of amino acids that are next to each amino acid. As much as the contact number is lower, it shows that our amino acids are

Method number	Thereshold	Max score	The best candidate peptide sequences
1	**0.359**	**2.777 for 477 position**	**Scattered areas**
2	**0.991**	**1.511 for 437 position**	**434 to 440**
3	**1**	**5.323 for 371 position**	**369 to 374**
4	**1.007**	**1.141 for 507 and 508 positions**	**504 to 511**
5	**1.016**	**1.22 for 43 position**	**40 to 46**
6	**2.101**	**8.686 for 492 position**	**489 to 495**

Table 12. Conclusion results for B cell linear epitope prediction from NP protein sequence.

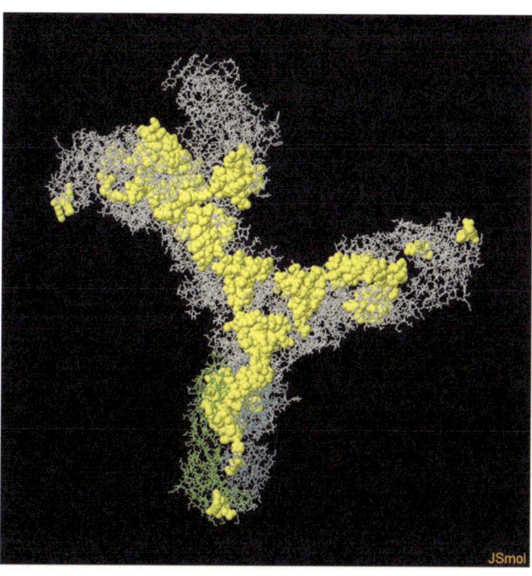

Figure 1. Discontinuous epitope(s) number one 3D structure for GP.

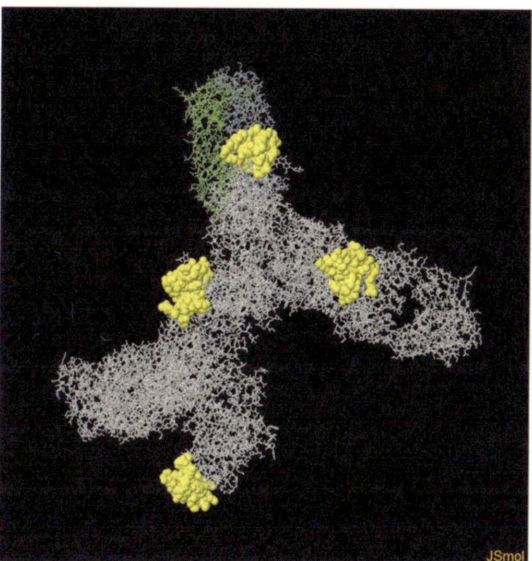

Figure 2. Discontinuous epitope(s) number two 3D structure for GP.

more external. For example, in this study, VAL residue in the 310th position had the lowest contact number and was more external.

C. Using IgPred web tool, epitope mapping of the amino acid sequence of the GP of ZEBOV was performed. The sequence of the chain "I" of glycoprotein was selected to study the interaction with the antibodies IgG, IgE and IgA.

As illustrated in **Table 15**, different regions in this chain had a different score for IgG but the end of the sequence had the highest score and was approximately 255–310 regions.

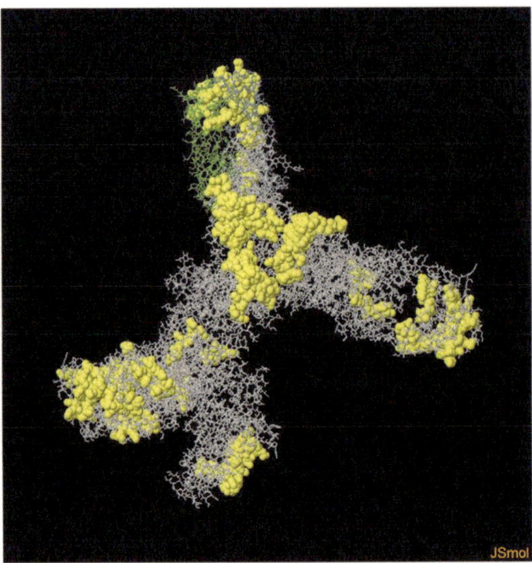

Figure 3. Discontinuous epitope(s) number three 3D structure for GP.

No.	Residues	Number of residues	Score
1	t:R31, t:S32, t:I33, t:P34, t:L35, t:G36, t:V37, t:I38, t:H39, t:N40, t:S41, t:V42, t:L43, t:Q44, t:V45, t:S46, t:D47, t:V48, t:D49, t:K50, t:L51, t:V52, t:C53, t:R54, t:D55, t:K56, t:L57, t:S58, t:S59, t:T60, t:N61, t:Q62, t:L63, t:R64, t:E100, t:I185, t:L186, t:P187, t:Q188, t:A189	40	0.796
2	t:E235, t:V236, t:D237, t:N238, t:L239, t:T240, t:Q243, t:F248, t:F252, t:Q255, t:L256, t:N257, t:E258, t:T259, t:I260, t:Y261, t:T262, t:S263, t:G264, t:K265, t:R266, t:S267, t:N268, t:T269, t:T270, t:G271, t:K272, t:L273, t:I274, t:W275, t:K276, t:V277, t:N278, t:R299, t:K300, t:I301, t:R302, t:S303, t:E304, t:E305, t:L306, t:S307, t:F308, t:T309, t:V310	45	0.701
3	t:R89, t:S90, t:G91, t:V92, t:P93, t:P94, t:K95, t:K114, t:K115, t:P116, t:D117, t:G118, t:S119, t:E120, t:C121, t:L122, t:P123, t:A124, t:A125, t:P126, t:D127, t:G128, t:I129, t:R130, t:G131, t:G145, t:P146, t:C147, t:A148, t:G149, t:D150, t:F151, t:D163, t:V169, t:Y171, t:R172, t:G173, t:G224, t:G226, t:T227, t:N228, t:E229, t:V230	43	0.614

Table 13. Three top scores predicted discontinuous epitope for B cell according to ElliPro web server.

Therefore, according to prediction with these web tools, 255–310 regions are the proper candidate for being epitope [4].

4.2.2.2. *Discontinuous epitopes prediction for NP*

No specific structure was found in the PDB for NP. Therefore, homology modeling of these types of proteins is needed to determine their structures. This goal has been achieved with the help of homology modeling, and its stages have been described below:

A. **Template Selection:** The template to be used in homology modeling should be based on target-template alignment, and the template that most closely resembles with our protein is selected as the template.

B. **Model Building:** The models are constructed using target-template alignment by ProMod3, and the areas that are shared and conserved between the template and the target are copied from the template on the model. The areas that are added or removed are rebuilt using fragment library. Side chains are then remodeled. Finally, its overall geometry is determined, but the loop areas are building with PROMOD-II.

Chain ID	Residue ID	Residue Name	Contact number	Disco Tope Score
I	261	TYR	13	-7.447
I	262	THR	11	-5.510
I	263	SER	11	-5.510
I	264	GLY	9	-2.952
I	265	LYS	15	-6.372
I	267	SER	15	-7.348
I	268	ASN	10	-3.780
I	269	THR	10	-4.544
I	270	THR	8	-3.423
I	271	GLY	13	-6.966
I	277	VAL	10	-5.284
I	278	ASN	8	-3.870
I	299	ARG	7	-2.755
I	300	LYS	5	-0.883
I	301	ILE	11	-5.702
I	302	ARG	8	-2.804
I	303	SER	12	-4.847
I	304	GLU	10	-4.323
I	306	LEU	11	-7.687
I	309	THR	7	-6.575
I	310	VAL	3	-3.807

Table 14. The predicted epitope sequence from 261 to 310 discontinuous amino acids for B cell according to DiscoTope web server.

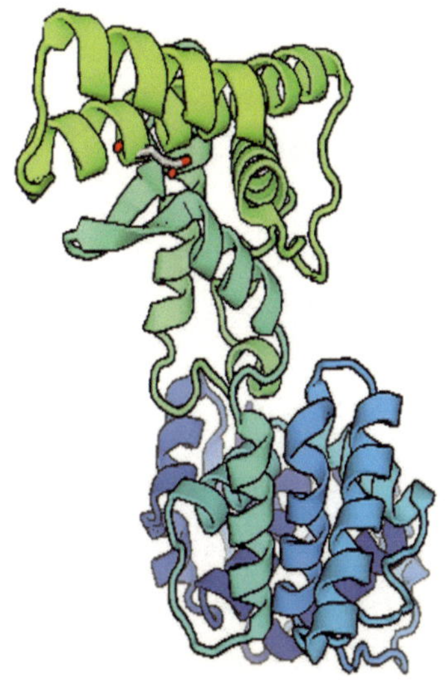

Figure 4. The model 3D structure.

Peptide seq	IgG Score	IgE Score	IgA Score	Prediction
YPYDVPDYAIEGRGARSIPLGVIHNSVLQ VSDVDKLVCRDKLSSTNQLRSVGLNLEG NGVATDVPSATKRWGFRSGVPPK	1.215	-0.148	0.546	IgG Epitope
VVNYEAGEWAENCYNLEIKKPDGSECLP AAPDGIRGFPRCRYVHKVSGTGPCAGDF AFHKEGAFFLYDRLASTVIYRGTT	1.019	-0.029	0.380	IgG Epitope
FAEGVVAFLILPQAKKDFFSSHPLREPVN ATEDPSSGYYSTTIRYQATGFGTNEVEYL FEVDNLTYVQLESRFTPQFLLQ	0.921	-0.238	0.459	IgG Epitope
LNETIYTSGKRSNTTGKLIWKVNPEIDTT IGEWAFWETKKNLTRKIRSEELSFTVVT HHQDTGEESASSGKLGLITNTIA	1.518	-0.844	1.009	IgG Epitope
GVAGLITGGRRTRR	0.691	-0.912	-0.156	Non-Epitope

Table 15. Antibodies scores for sequence of "I" chain of GP according to IgPred web server.

C. Results: The SWISS-MODEL template library (SMTL version 2017-09-21, PDB release 2017-09-15) was searched with BLAST and HHBlits for evolutionarily related structures matching the target sequence is presented in **Table 16** and the model 3D structure is also shown in **Figure 4**.

D. After building homology modeling, ElliPro web tool was performed to predict conformational epitopes of NP. These results are shown in **Table 17** and (**Figures 5** and **6**).

Template	Seq Identity	Oligo-state	Found by	Method	Resolution	Seq Similarity	Range	Coverage	Description
4ypi.2.A	100.00	hetero-oligomer	HHblits	X-ray	3.71Å	0.60	39 - 384	0.47	Nucleoprotein

Table 16. Details on the template search.

No.	Residues	Number of residues	Score
1	A:H265, A:P266, A:L267, A:A268, A:R269, A:T270, A:A271, A:K272, A:V273, A:K274, A:N275, A:E276, A:V277, A:N278, A:S279, A:K281, A:A282, A:S285, A:K289, A:H290, A:G311, A:F313, A:P314, A:Q315, A:G333, A:V334, A:N335, A:V336, A:G337, A:E338, A:Q339, A:Y340, A:Q341, A:Q342, A:L343, A:R344, A:E345, A:A346, A:A347, A:T348, A:E349, A:A350, A:K352, A:Q353, A:L354, A:Q355, A:Q356, A:Y357, A:A358, A:E359, A:S360, A:R361, A:E362, A:L363, A:D364, A:H365, A:L366, A:G367, A:L368, A:D369, A:D370, A:Q371, A:E372, A:K373, A:K374, A:I375, A:L376, A:M377, A:N378, A:F379, A:H380, A:Q381, A:K382, A:K383, A:N384	75	0.752
2	A:T258, A:E259, A:R260, A:G261	4	0.75

Table 17. Top scores predicted discontinuous epitope for B cell according to ElliPro web server.

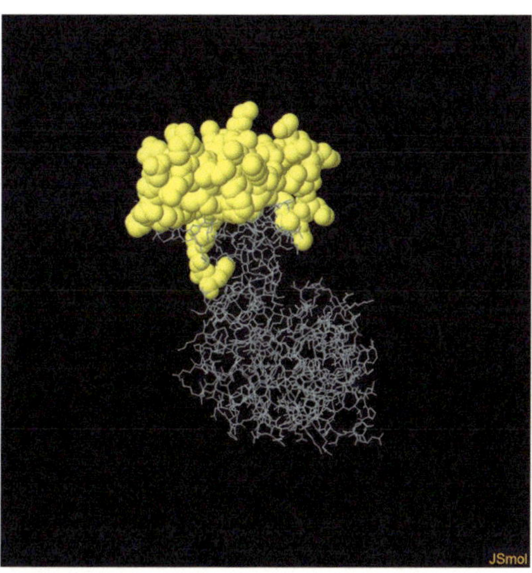

Figure 5. Discontinuous epitope(s) number one 3D structure for NP.

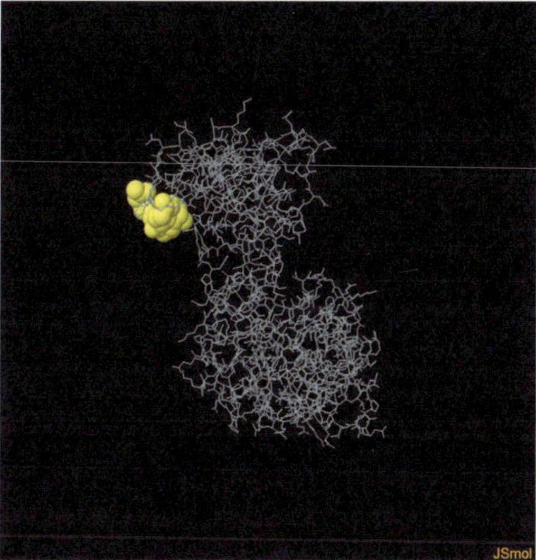

Figure 6. Discontinuous epitope(s) number two 3D structure for NP.

5. Conclusion

The hemorrhagic fever (HF) is a lethal disease from ZEBOV that caused the death of many people in Africa. The glycoprotein of ZEBOV is the only protein on the surface of the virus, and has severe cytotoxicity effects, also nucleoprotein can stimulate the immune response, stronger than GP, therefore, it was offered to design the vaccine against both of them [4, 13,

16, 17].Vaccination is a good idea to prevent infection and limit the spreading of Ebola virus. So far, several vaccines against Ebola glycoprotein have been tested on animals, but no licensed vaccine has been reported in humans [31, 32]. There are several types of vaccine, in which one of them is a peptide vaccine, and it is the goal of this analysis. These vaccines are designed based on the epitopes fragment of antigens and they are safer and easy to prepare. For T cells, these epitopes are linear and they must bind to MHC I and MHC II at first, but for B cells, 3D conformation of antigens is very important so it can be discontinuous epitopes [6]. These discontinuous epitopes may be near to each other in the 3D conformation but far from each other in the amino acid sequence or first structure [4, 10]. The experimental work on this virus is very difficult and biosafety level is 4. With the help of immunoinformatics tools for vaccine design, the use of laboratory work is reduced, and hence we save more expense and time. Immunoinformatics approach can help us to predict T cell and B cell epitopes, and also have application in *in silico* vaccination. By predicting T cell and B cell epitopes, we can find the peptide that is useful for *in silico* vaccination or for designing multipeptide vaccine [11]. By using ProPred, CTLPred, SYFPEITHI and ProPred1 web tools for predicting the epitopes for T cells, we had been able to introduce peptides that can be the candidate for designing multipeptide vaccines. With the use of ElliPro, DiscoTope, IgPred web servers and linear epitope prediction methods from IEDB predicting epitopes for B cells, and these peptides can induce the immune response and design for the peptide-based vaccine. With the help of immunoinformatics tools for predicting epitopes for T cells and B cells, we can design the multipeptide vaccine, and this vaccine can include both epitopes from GP and NP, which is useful for increasing immune response against Ebola virus. As conclusion for linear epitopes that bind to MHC I "TRKIRSEEL," "SRFTPQFLL" and "ETTQALQLF" peptides for GP [4], and "RLHPLARTA," "SRELDHLGL" and "VKNEVNSFK" peptides for NP are candidate epitopes. Also for MHC class II: "IILFQRTFSIPLGVI," "CRYVHKVSGTGPCAG" and "FFLYDTLAS" peptides for GP and "FLSFASLFL," "NRFVTLDGQQFYWPV" and "FRLMRTNFL" peptides for NP have the highest scores and the most frequent within this analysis. For B cell linear epitope prediction, results are shown in **Tables 11** and **12**. As a final result for conformational epitopes, we can only say, peptide sequence from 255 to 310 amino acids for GP has a higher score. These peptides are able to be the candidate for the vaccine against Ebola virus. It should be noted that these results just in *in silico* are valid and need laboratory (in vivo and in vitro) confirmation.

Author details

Maryam Hemmati[1†], Ehsan Raoufi[1*†] and Hossein Fallahi[2]

*Address all correspondence to: ehsan.raoufi@gmail.com

1 Department of Medical Biotechnology, School of Allied medicine, Iran University of Medical Sciences, Tehran, Iran

2 Department of Biology, Faculty of Science, Razi University, Kermanshah, Iran

† These authors contributed equally.

References

[1] Lee JE et al. Structure of the Ebola virus glycoprotein bound to a human survivor antibody. Nature. 2008;**454**(7201):177

[2] Saeed MF et al. Cellular entry of Ebola virus involves uptake by a macropinocytosis-like mechanism and subsequent trafficking through early and late endosomes. PLoS Pathogens. 2010;**6**(9):e1001110

[3] Gire SK et al. Genomic surveillance elucidates Ebola virus origin and transmission during the 2014 outbreak. Science. 2014;**345**(6202):1369-1372

[4] Raoufi E, Hemmati M, EinAbadi H, Fallahi H. Predicting candidate epitopes on Ebolaviruse for possible vaccine development. In: Proceedings of the 2015 IEEE/ACM International Conference on Advances in Social Networks Analysis and Mining 2015. New York, NY, USA: ACM; Aug 2015. pp. 1083-1088

[5] Chakraborty S. Ebola vaccine: Multiple peptide-epitope loaded vaccine formulation from proteome using reverse vaccinology approach. International Journal of Pharmacy and Pharmaceutical Sciences. 2014;**6**:407-412

[6] Patronov A, Doytchinova I. T-cell epitope vaccine design by immunoinformatics. Open Biology. 2013;**3**(1):120139

[7] Martin JE et al. A DNA vaccine for Ebola virus is safe and immunogenic in a phase I clinical trial. Clinical and Vaccine Immunology. 2006;**13**(11):1267-1277

[8] Korber B, LaBute M, Yusim K. Immunoinformatics comes of age. PLoS Computational Biology. 2006;**2**(6):e71

[9] Moise L, De Groot AS. Putting immunoinformatics to the test. Nature Biotechnology. 2006;**24**(7):791-792

[10] Sharon J et al. Discovery of protective B-cell epitopes for development of antimicrobial vaccines and antibody therapeutics. Immunology. 2014;**142**(1):1-23

[11] Tomar N, De Rajat K. Immunoinformatics: An integrated scenario. Immunology. 2010; **131**(2):153-168

[12] Dowling W et al. Influences of glycosylation on antigenicity, immunogenicity, and protective efficacy of ebola virus GP DNA vaccines. Journal of Virology. 2007;**81**(4):1821-1837

[13] Qiu X et al. Characterization of *Zaire ebolavirus* glycoprotein-specific monoclonal antibodies. Clinical Immunology. 2011;**141**(2):218-227

[14] Volchkova VA et al. Genomic RNA editing and its impact on Ebola virus adaptation during serial passages in cell culture and infection of guinea pigs. The Journal of Infectious Diseases. 2011;**204**(Suppl_3):S941-S946

[15] Ali MT, Islam MO. A highly conserved GEQYQQLR epitope has been identified in the nucleoprotein of Ebola virus by using an in silico approach. Advances in Bioinformatics. Feb 1 2015;**2015**

[16] McElroy AK et al. Human Ebola virus infection results in substantial immune activation. Proceedings of the National Academy of Sciences. 2015;**112**(15):4719-4724

[17] Patronov A, Doytchinova I. T-cell epitope vaccine design by immunoinformatics. Open Biology. Jan 1 2013;**3**(1):120139

[18] Sundar K, Boesen A, Coico R. Computational prediction and identification of HLA-A2. 1-specific Ebola virus CTL epitopes. Virology. 2007;**360**(2):257-263

[19] Haste Andersen P, Nielsen M, Lund O. Prediction of residues in discontinuous B-cell epitopes using protein 3D structures. Protein Science. 2006;**15**(11):2558-2567

[20] Li Pira G, Ivaldi F, Moretti P, Manca F. High throughput T epitope mapping and vaccine development. BioMed Research International. Jun 15 2010;**2010**

[21] Kringelum JV et al. Reliable B cell epitope predictions: Impacts of method development and improved benchmarking. PLoS Computational Biology. 2012;**8**(12):e1002829

[22] Tsurui H, Takahashi T. Prediction of T-cell epitope. Journal of Pharmacological Sciences. 2007;**105**(4):299-316

[23] Tong JC, Tan TW, Ranganathan S. Methods and protocols for prediction of immunogenic epitopes. Briefings in Bioinformatics. 2006;**8**(2):96-108

[24] Khan M et al. Epitope-based peptide vaccine design and target site depiction against Ebola viruses: An immunoinformatics study. Scandinavian Journal of Immunology. 2015;**82**(1):25-34

[25] Lapelosa M et al. In silico vaccine design based on molecular simulations of rhinovirus chimeras presenting HIV-1 gp41 epitopes. Journal of Molecular Biology. 2009;**385**(2):675-691

[26] Larsen JE, Lund O, Nielsen M. Improved method for predicting linear B-cell epitopes. Immunome Research. 2006;**2**(1):2

[27] Emini EA et al. Induction of hepatitis a virus-neutralizing antibody by a virus-specific synthetic peptide. Journal of Virology. 1985;**55**(3):836-839

[28] Karplus P, Schulz G. Prediction of chain flexibility in proteins. Naturwissenschaften. 1985;**72**(4):212-213

[29] Kolaskar A, Tongaonkar PC. A semi-empirical method for prediction of antigenic determinants on protein antigens. FEBS Letters. 1990;**276**(1–2):172-174

[30] Parker J, Guo D, Hodges R. New hydrophilicity scale derived from high-performance liquid chromatography peptide retention data: Correlation of predicted surface residues with antigenicity and X-ray-derived accessible sites. Biochemistry. 1986;**25**(19):5425-5432

[31] Wu S et al. Prediction and identification of mouse cytotoxic T lymphocyte epitopes in Ebola virus glycoproteins. Virology Journal. 2012;**9**(1):111.4

[32] Atanas P, Irini D. T-cell epitope vaccine design by immunoinformatics. Open Biology 2013;**3**:120139

Chapter 5

Bioinformatics Approach to Screening and Developing Drug against Ebola

Usman Sumo Friend Tambunan,
Ahmad Husein Alkaff and
Mochammad Arfin Fardiansyah Nasution

Additional information is available at the end of the chapter

http://dx.doi.org/10.5772/intechopen.72278

Abstract

Ebola is an acute disease causing hemorrhagic fever marked with high mortality rate. The patients who suffer from Ebola only receive palliative care because there is no available drug which can consistently cure this disease. To date, no cure has been found to treat this disease. Bioinformatics and computer-aided drug discovery and development (CADDD) are employed to utilize the readily available genomic and proteomic data and enhance the hit rate of the novel and repurposed drug for Ebola therapy. Additionally, the time and cost of wet laboratory experiments can be drastically reduced by the support of bioinformatics approach. Our laboratory has succeeded not only in creating the bioinformatics research pipeline but also screening and developing the drug candidate to cure Ebola. Through pharmacophore-based virtual screening and molecular docking simulations, we discovered that about three Indonesian natural product compounds have noteworthy molecular interactions against EBOV VP35 protein, which are responsible for the RNA synthesis of the Ebola. These compounds can be reevaluated further through advances in in silico simulation and in vitro experiments.

Keywords: Ebola, *Ebola virus*, bioinformatics, CADDD, molecular docking

1. Introduction

Ebola, previously known as Ebola virus disease, is an acute viral infection causing hemorrhagic fever marked by high mortality rate in human and nonhuman primates [1]. It is a zoonotic disease transmitted by direct contact with mucosal tissue or bodily fluids (blood, feces, and other secreted fluids) of the infected living or dead human and animal [2–4]. The animal reservoir for this disease is still unknown. Fruit bat (*Hypsignathus monstrosus, Epomops*

franqueti, and *Myonycteris torquata*) which belongs to Pteropodidae family is suspected as the most likely host of Ebola, although the linkage has not been confirmed [5–7].

Ebola is an enveloped, nonsegmented, negative-sense, single-stranded RNA virus which belongs to *Ebolavirus* genus, *Filoviridae* family, and *Mononegavirales* order [8, 9]. Ebola virus (EBOV), Tai Forest virus (TAFV), Reston virus (RESTV), Sudan virus (SUDV), and Bundibugyo virus (BDBV) are the virus making up the *Ebolavirus* genus [10]. EBOV and SUDV come as the most frequent outbreak-causing virus which has the case-fatality rate of 76 and 55% (CI 95%), respectively [11]. On the other hand, RESTV causes death in primates such as gorillas and chimpanzees but not known to have caused disease in humans [12, 13].

The Ebola virus genomic RNA is consisted of around 19,000 nucleotides [14]. It encodes seven structural protein, namely, nucleoprotein (NP), glycoprotein (GP), RNA-dependent RNA polymerase (L), matrix protein (VP40), and three nucleocapsid proteins (VP24, VP30, and VP35) [15, 16]. It also encodes one nonstructural protein, the secretory glycoprotein (sGP) [17]. The genome is linearly arranged as follows: 3′-leader-NP-VP35-VP40-GP/sGP-VP30-VP24-L-trailer-5′ [14, 17, 18].

The seven structural proteins and one nonstructural protein have an imperative role in Ebola virus life cycle [16]. NP: viral replication and scaffold for additional viral proteins. GP: binds to receptors on the cell surface and membrane fusion, pathogenicity. sGP: inhibits neutrophil function and adsorbs neutralizing antibodies. L: synthesis of positive-sense RNA. VP40: viral assembly and budding, structural integrity of viral particles, and maturation of the virion. VP24: nucleocapsid formation, encapsulates and shields viral genome from nucleases, viral replication. VP30: viral transcription activator. VP35: multi-virulence functionality, innate immune antagonist, and an RNAi silencing suppressor [16, 17].

The patient who suffers from Ebola shows no symptoms during the initial infection. After the incubation for about 4–10 days, the general symptoms such as fever, myalgia, and malaise and sometimes accompanied by chills appear. These symptoms often confused with dengue or malaria in tropical climates [3, 19]. As the infection progresses, the patient shows flu-like symptoms accompanied by gastrointestinal symptoms. In severe cases, Ebola developed into a conjunctival hemorrhage, epistaxis, melena, hematemesis, coagulation abnormalities, and a range of hematological irregularities. The neurological symptoms such as encephalopathy, convulsions, and delirium may also occur during the late stage of the infection [19, 20]. The patient dies around 6–9 weeks after the symptoms appear [21]. With the nonspecific symptoms, severe morbidity, and high mortality rate, the World Health Organization (WHO) has acknowledged Ebola as one of the most malignant diseases in the world [22].

The first recorded Ebola outbreak emerged in Sudan between June and November 1976. It mainly affected Nzara, Maridi, Tembura, and Juba where 150 of 284 victims died (the mortality rate of 53%) [2, 23]. After the first outbreak, 19 other outbreaks have occurred in Africa with the mean fatality rate of 65.4% [11].

The last and the most extensive Ebola outbreak was announced by the WHO on March 23, 2014. This outbreak appears to have emerged in the Guéckédou district of the southeast region of Republic of Guinea [24–26]. The WHO announced the epidemic to be a Public

Health Emergency of International Concern (PHEIC) on August 8, 2014, due to the severe consequences if Ebola ever spread around the globe. PHEIC was disclosed because of the weak health services of Guinea, Liberia, Sierra Leone, and other neighboring countries at risk in combating Ebola and the continuing transmission with a high fatality rate of Ebola in West Africa [26]. When the outbreak ends in March 2016, Ebola has claimed 1310 lives out of 28,616 reported cases [27, 28]. Even though the damage caused by the last outbreak of Ebola is calamitous, there is still no FDA-approved antiviral drug to treat this disease.

Ebola is considered as one of the neglected tropical diseases because the outbreaks take place in the poor populations with limited resources, mostly in West Africa [29]. The research and drug development for Ebola have been neglected for decades because the drug developers regard it as a commercially unattractive project to invest their resource. The negligence occurs to all tropical diseases by only 13 out of 1393 new approved drugs between 1975 and 1999 that were indicated for tropical disease [30]. However, the frequent outbreaks in the last decade and the massive outbreak which was occurred in 2014 have drawn much attention to drug development for Ebola [16]. Without available treatment or vaccine, paramedic only relied on palliative care for the infected patients and barrier methods to prevent the transmission [31]. Hence, the researchers investigating ways for helping people just infected with Ebola (treatment) and preventing people to get infected when exposed to Ebola (vaccine) [32].

The conventional medical treatment for Ebola is a supportive care with intravenous fluids or oral rehydration with electrolyte solutions. The reason being that the virus interferes with blood clotting and disrupts electrolyte balance. Thus, such intervention can help to keep up the condition of the patient. However, such intervention is not enough for severely ill patients to sustain and recover [21, 32].

Zmapp, a combined humanized monoclonal antibody, was tested as a passive immunotherapy against Ebola. The preclinical test was conducted by Mapp Biopharmaceutical. This monoclonal antibody shows 100% efficacy in preventing lethal disease on cynomolgus macaques when treatment is initiated up to 5 days postinfection of EBOV [31].

Other experimental therapies developed a novel synthetic adenosine analog, BCX4430. This compound shows in vitro and in vivo activity by inhibiting viral RNA polymerase function, acting as a non-obligate RNA chain terminator. BCX4430 protects both mice and guinea pig models from a severe infection of Ebola virus and Marburg virus. In addition, this compound completely protects cynomolgus macaques from Marburg virus infection if administered as late as 48 h after infection [33].

Not only does the research focus on the development of a novel drug, but the research is also conducted to identify potential repurposed therapeutic agents for the treatment of Ebola [34, 35]. Toremifene and clomiphene, the selective estrogen reuptake modulators, are currently known as the drug to treat breast cancer and infertility, respectively. Both drugs inhibit Ebola virus entry into the cell by preventing the late stage membrane fusion. These drugs show an inhibition activity by more than 90% in vitro. Higher dose than the standard clinical range is needed to achieve the therapeutic effect on Ebola. However, a higher dose would increase the risk of serious side effect of toremifene and clomiphene, which are electrolyte derangements and ocular adverse effect, respectively [36].

Other experiments screen amiodarone, a multichannel ion blocker; sertraline, selective serotonin reuptake inhibitor; and bepridil, a calcium channel blocker as a repurposed therapeutic agent targeting Ebola. Amiodarone works by the induction of Niemann-Pick C-like phenotype that inhibits late endosomal entry of Ebola virus [37]. Sertraline and bepridil work in a similar fashion to amiodarone. Both drugs show inhibition activity in an in vitro test by more than 90% [35].

Several vaccines have also been developed to prevent the Ebola. ChAd3-ZEBOV, which has developed by GlaxoSmithKline in collaboration with the US National Institute of Allergy and Infectious Diseases, is a chimpanzee-derived adenovirus vector with an Ebola virus gene inserted. This vaccine induced uniform protection against acute lethal Ebola virus in cynomolgus macaques. However, the protection of this vaccine declines over several months [38].

The other vaccine, which is developed by the Public Health Agency of Canada in Winnipeg, is rVSV-ZEBOV. It uses an attenuated vesicular stomatitis virus which has been genetically modified to express glycoprotein of Ebola virus. The rVSV-ZEBOV has undergone a ring vaccination phase 3 efficacy trial which assesses the protective activity of rVSV-ZEBOV against Ebola virus in human beings. The result shows that rVSV-ZEBOV offers substantial protection against Ebola virus infection. Both randomized and a non-randomized clusters of vaccinated individuals show no disease development from the challenge performed 10 days postvaccination [39].

The Center for Disease Control and Prevention considered Ebola virus as a tier 1 select agent because it possesses a considerable risk of intentional misuse with a severe threat to public health and safety [40]. Researchers need to fill the APHIS/CDC Form 1 in order to register for possession, use, and transfer of Ebola virus. All requirements including the availability of Biosafety Level (BSL) 4 laboratory and certified personnel are needed to get access to Ebola virus sample [41]. Thus, to get a suitable sample, researchers tend to move their experiments on the genetically modified virus that can express part of known Ebola virus genome because it is not subjected to select agent [42].

Genomic and proteomic data of Ebola virus has been collected each time the outbreak occurred and stored in the open source database. Also, the Ebola virus protein interaction with the corresponding drug lead through in vitro test has also been increased in the past decades. To date, the protein three-dimensional (3D) structure of Ebola virus NP, VP35, VP40, GP, VP30, and VP24 has been available in Protein Databank (PDB). In addition, the active site residues of each protein have also been identified, except for NP. L is the only Ebola virus protein with unavailable 3D structure and unidentified active site [16]. Thus, researchers use a bioinformatics approach to utilize the readily available genomic and proteomic data to research drug design and discovery.

Computer-aided drug discovery and development (CADDD) is employed to accelerate hit identification, hit-to-lead selection, enhance absorption, distribution, metabolism, excretion, and toxicity (ADMET) profile and avoid another safety issue [43]. This approach is currently growing and adapted quickly by pharmaceutical industry and academia because it reduces the time and cost of drug research [44, 45]. Currently, 16 compounds (Aliskiren, Boceprevir, Captopril, Dorzolamide, Indinavir, LY-517717, Nolatrexed, NVP-AUY922, Oseltamivir,

Raltegravir, Ritonavir, Rupintrivir, Saquinavir, TMI-005, Tirofiban, and Zanamivir) are in clinical trial or have been approved for therapeutic use. These compounds are the examples of successful application of CADDD [46, 47]. Through CADDD, the hit rate of the novel and repurposed drug for Ebola therapy could be improved.

A consistently effective treatment for Ebola is currently not yet available. Present therapeutic options are directed at palliative and supportive care to maintain and prolong the patient life. The majority of treatment, novel or repurposed drug, have been developed, but none of them are entirely satisfactory. In attempts to find a drug in the treatment of Ebola, inhibitors targeting EBOV VP35 have received little attention even though it has a critical function in host immune evasion and viral RNA synthesis. Our objective is to find the optimal in silico Ebola therapeutic agents which later will be implemented in the wet laboratory.

2. Our in silico method

In this chapter, we will discuss the result of our in silico approach against EBOV VP35, one of the viral protein of EBOV which is responsible for the viral RNA synthesis and as the RNAi silencing suppressor agents [48, 49]. Moreover, this protein was also being studied by Brown et al. in 2014, which discovered the actual pose of their selected inhibitors against the EBOV VP35 in their perspective binding site and also deposited their work in RCSB Protein Databank (PDB) through several PDB IDs [50]. Thus, their proteins can be used as the template for pharmacophore mapping model for our docking simulation approach. Moreover, we also deployed the Indonesian natural product compounds for virtual screening purpose to find the suitable lead compounds for combating Ebola. The reason for choosing the Indonesian natural product compounds because of Indonesia, as one of the largest megadiversity countries, has no less than 38,000 flowering plants that grow around the nation, with 55% of them are endemic plants [51, 52].

First, we prepared the Indonesian natural product compounds by searching the molecular structures through several journals and databases [53–69], after which we were drawing them using ChemBioDraw 14.0 software. From this step, we obtained about 3429 compounds in the process. All of these ligands were then protonated, washed, and minimized by using MOE 2014.09 software [70]. These ligands were saved for the docking simulation purpose. For the EBOV VP35 protein, we selected the PDB ID: 4IBC as our protein, and we determined the pharmacophore site through standard protein-ligand interaction fingerprints (PLIF) protocol of MOE 2014.09 software. This step generated the pharmacophore model around the binding site of EBOV VP35 after we performed the protonating process of EBOV VP35 through "LigX" feature of MOE 2014.09 software. Later on, we conducted molecular docking simulation using the modification of our current approach [71, 72]. Instead of using "Triangle Matcher" and "London dG," we used "Pharmacophore" and "Affinity dG" for "Placement" and "Rescoring 1" parameters to accommodate the pharmacophore model that created in an earlier phase, while the rest of parameters were set according to the default setup. First, the STD1 ligand (IUPAC name: 2-(4-(4-(2-chlorobenzoyl)-5-(2-chlorophenyl)-2,3-dioxo-2,3-dihydro-1H-pyrrol-1-yl) phenyl)acetic acid) and 100 decoy ligands were docked into the binding site to validate the pharmacophore model. "Rigid Receptor" and "Induced-Fit" protocols were performed against the Indonesian natural product compounds and STD1 ligand later on.

In an attempt of searching the proper pharmacophore site in the binding site of EBOV VP35, we utilized the PLIF protocol from MOE 2014.09 software by using STD1 ligand as the template. From this approach, we figured out that the binding site of EBOV VP35 protein consists of three pharmacophore sites, as it displayed in **Figure 1**. One hydrophobic spot is affiliated with Lys248 residues through arene-hydrogen interaction, while two H-bond acceptors, lone-pair sites, are connected with Gln241 and Lys251. These sites were responsible for the binding attachment of the STD01 ligand when bound to EBOV VP35 protein. Thus, it can be predicted that any ligands that bind to these residues may exhibit the same antiviral activities like STD01 ligand.

The pharmacophore sites were later validated by having the STD01 ligand and 100 decoys to be screened through molecular docking simulation. In this phase, we deployed "virtual screening" approach as our docking protocol, with pharmacophore model that was included in the simulation. After the screening was conducted, we discovered that all of the decoys did not pass the test, indicating this method was validated and did not create the "false-positive" ligand that may result during docking simulation. Furthermore, the STD01 ligand passed this test with a $\Delta G_{binding}$ score of −5.2778 kcal/mol and RMSD value of 1.5487 Å. This result was shown that the parameters that were set earlier were decent enough to be reproduced in the next simulation. The comparison of the initial and screened poses of STD01 ligand is shown in **Figure 2**.

The pharmacophore-based docking simulation of EBOV VP35 protein was later performed against the 3429 Indonesian natural product ligands that were already prepared. From the simulations, we acquired 20 ligands that matched with the pharmacophore model of EBOV VP35, which means that other 3409 ligands did not possess the properties that needed to pass the initial pharmacophore screening. In the first docking simulation (Rigid Docking protocol),

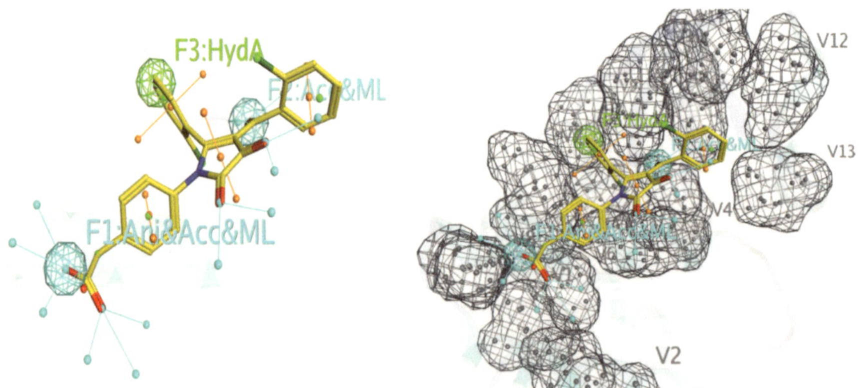

Figure 1. The pharmacophore model of the STD01 ligand in the binding site of EBOV VP35 protein. According to the PLIF feature of MOE 2014.09 software, the STD01 ligand comprises three pharmacophore sites: one hydrophobic point and two acceptor/lone-pair points (left). In the docking simulations, we deployed the "exclude points" to indicate the residues that exist in the VP35 binding site and prevent the larger ligands to interact with the binding site.

we found four Indonesian natural product ligands, namely, multifloroside, myricetin 3-robinobioside, kaempferol 3-(6G-malonylneohesperidoside), and theasaponin, which have the $\Delta G_{binding}$ score lower than the STD01 ligand. The molecular structures of these ligands can be seen in **Figure 3**.

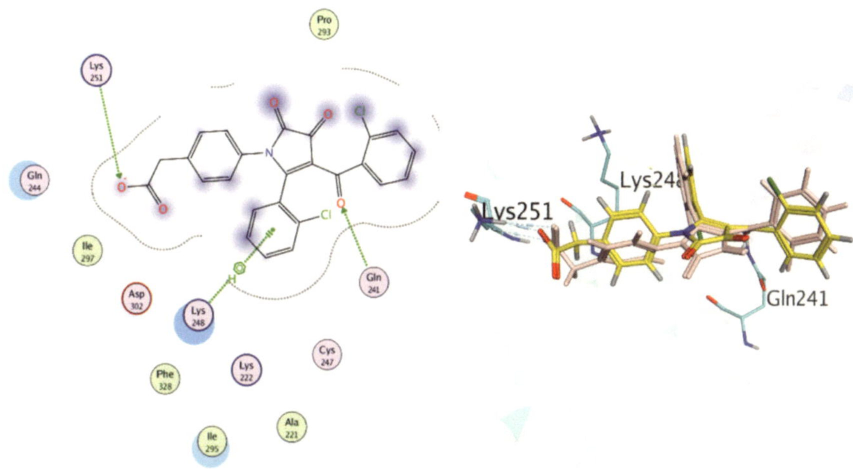

Figure 2. The binding interaction of the STD01 ligand and EBOV VP35 binding site. The 2D interaction after docking simulation is displayed in the left figure, while the right figure presents the difference between the initial pose (shown in yellow) and after the docking simulation was conducted (shown in pink).

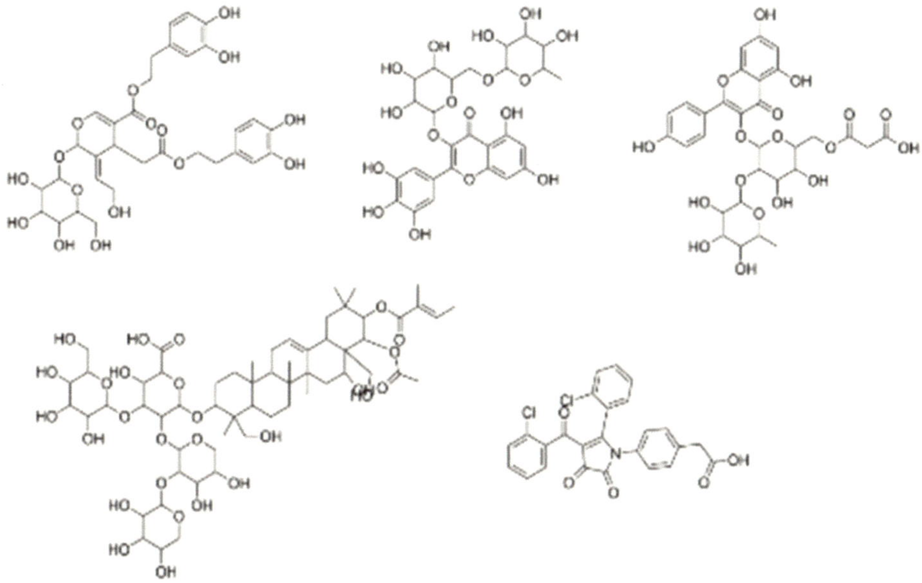

Figure 3. The molecular structure of multifloroside (top left), myricetin 3-robinobioside (top middle), kaempferol 3-(6G-malonylneohesperidoside) (top right), theasaponin (bottom left), and 2-(4-(4-(2-chlorobenzoyl)-5-(2-chlorophenyl)-2,3-dioxo-2,3-dihydro-1H-pyrrol-1-yl)phenyl)acetic acid (bottom right).

Molecule name	$\Delta G_{binding}$ (RMSD)	H-bond interaction residues
Multifloroside	−10.8405 kcal/mol (3.2691)	Arg225, Tyr229, Lys 248, and Lys251
Myricetin 3-robinobioside	−10.0897 kcal/mol (1.2275)	Lys222, Arg225, Gln241, and Lys251
Kaempferol 3-(6G-malonylneohesperidoside)	-9.8721 kcal/mol (1.0311)	Gln241, Gln244, Lys251, and His296
Theasaponin	-9.0175 kcal/mol (0.4352)	Arg225, Gln241, and Lys251
STD01 ligand (*standard*)	-8.4579 kcal/mol (0.7747)	Gln241, Lys248, and Lys251

Table 1. The results of molecular docking simulation.

After the first docking simulation had been performed, the second docking simulation (Induced-Fit protocol) was utilized against these four proteins to revalidate the docking pose that was produced in the previous simulations. If the RMSD difference was lower than 2.0 Å, it means that the docking pose is good enough and may be reproduced in the actual simulation [73]. In this simulation, we found that multifloroside ligand has the lowest $\Delta G_{binding}$ score of −10.8405 kcal/mol, followed by myricetin 3-robinobioside (−10.0897 kcal/mol), kaempferol 3-(6G-malonylneohesperidoside) (−9.8721 kcal/mol), and theasaponin (−9.0175 kcal/mol). These results were significantly lower than the STD01 ligand, which sits in −9.0175 kcal/mol. However, we must take into account that the RMSD value of multifloroside ligand was 3.2691 Å, higher than 2.0 Å; it means that the docking pose that was generated during the docking simulation was not acceptable. Meanwhile,

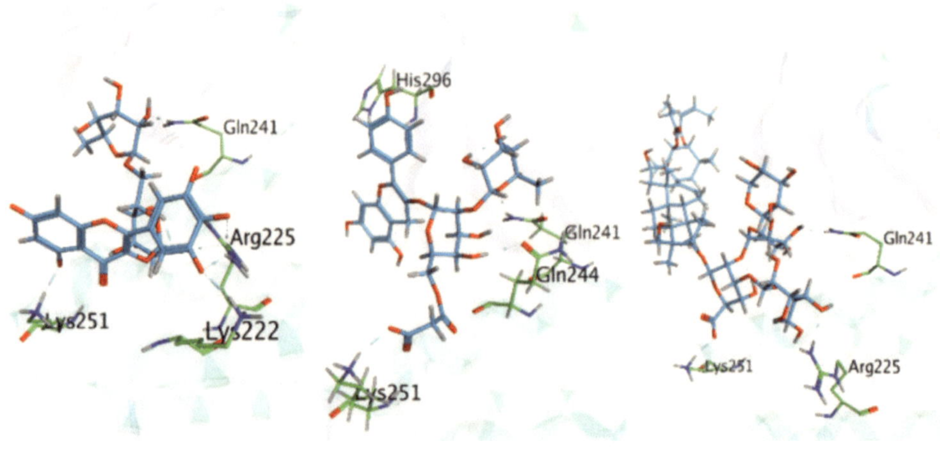

Figure 4. The interacting residues of EBOV VP35 protein with myricetin 3-robinobioside (left), kaempferol 3-(6G-malonylneohesperidoside) (middle), and theasaponin (right).

http://dx.doi.org/10.5772/intechopen.72278

the other three ligands possessed the tolerable RMSD value (1.2275, 1.0311, and 0.4352 Å for myricetin 3-robinobioside, kaempferol 3-(6G-malonylneohesperidoside), and theasaponin, respectively). Furthermore, we also observed the interactions between the ligands and the binding site of EBOV VP35. From the docking simulation, we figured out that all three remaining ligands made interactions with Gln241 and Lys251, which are important in suppressing the EBOV VP35 activity. The full results of molecular docking simulations can be seen in **Table 1** and **Figure 4**.

3. Conclusions

Without no doubt, the drug developments of Ebola are desperately needed due to high pathogenicity and mortality rate that emitted by this disease. Through this chapter, we present that bioinformatics and CADDD, especially the pharmacophore-based drug design, may be the solution to significantly increase the viability of the newly discovered lead compounds that can be introduced as the drug candidate of Ebola virus, which can be supported later through in vitro study to validate the results that we found in previous research. The dry lab experiments should play a significant role in the development of drugs, not only Ebola but also for all diseases due to low cost and not a time-consuming process. Therefore, the improvements and developments of bioinformatics and CADDD should also speed up the time that we needed to obtain the drug candidates for our health problems.

Acknowledgements

This book chapter is financially supported by Penelitian Unggulan Perguruan Tinggi (PUPT) 2017, granted by Ministry of Research, Technology, and Higher Education, the Republic of Indonesia through Directorate of Research and Community Engagements, Universitas Indonesia, with no: 2716/UN2.R3.1/HKP.05.00/2017.

Conflict of interest

None are declared.

Author details

Usman Sumo Friend Tambunan*, Ahmad Husein Alkaff and Mochammad Arfin Fardiansyah Nasution

*Address all correspondence to: usman@ui.ac.id

Bioinformatics Research Group, Department of Chemistry, Faculty of Mathematics and Natural Science, Universitas Indonesia, Depok, Indonesia

References

[1] Rajak H, Jain DK, Singh A, et al. Ebola virus disease: past, present and future. Asian Pacific Journal of Tropical Biomedicine. 2015;**5**:337-343

[2] Pourrut X, Kumulungui B, Wittmann T, et al. The natural history of Ebola virus in Africa. Microbes and Infection. 2005;**7**:1005-1014

[3] Ansari AA. Clinical features and pathobiology of Ebolavirus infection. Journal of Autoimmunity. 2015;**55**:1-9

[4] Bausch DG, Towner JS, Dowell SF, et al. Assessment of the risk of Ebola virus transmission from bodily fluids and fomites. The Journal of Infectious Diseases. 2007;**2699**:2-7

[5] Gatherer D. The 2014 Ebola virus disease outbreak in West Africa. The Journal of General Virology. 2014;**95**:1619-1624

[6] Smith I, Wang LF. Bats and their virome: An important source of emerging viruses capable of infecting humans. Current Opinion in Virology. 2013;**3**:84-91

[7] Feldmann H, Geisbert TW. Ebola haemorrhagic fever. Lancet. 2011;**377**:849-862

[8] Bishop BM. Potential and emerging treatment options for Ebola virus disease. The Annals of Pharmacotherapy. 2015;**49**:196-206

[9] Lawrence P, Danet N, Reynard O, et al. Human transmission of Ebola virus. Current Opinion in Virology. 2017;**22**:51-58

[10] Kuhn JH, Bao Y, Bavari S, et al. Virus nomenclature below the species level: A standardized nomenclature for laboratory animal-adapted strains and variants of viruses assigned to the family Filoviridae. Archives of Virology. 2013;**158**:1425-1432

[11] Lefebvre A, Fiet C, Belpois-Duchamp C, et al. Case fatality rates of Ebola virus diseases: A meta-analysis of World Health Organization data. Médecine et Maladies Infectieuses. 2014;**44**:8-12

[12] Formenty P. Ebola virus disease. In: Emerging Infectious Diseases: Clinical Case Studies. London: Elsevier, 2014. Epub ahead of print 2014. DOI: 10.1016/B978-0-12-416975-3.00009-1

[13] Kortepeter MG, Bausch DG, Bray M. Basic clinical and laboratory features of filoviral hemorrhagic fever. The Journal of Infectious Diseases. 2011;**204**(Suppl):S810-S816

[14] Ikegami T, Calaor AB, Miranda ME, et al. Genome structure of Ebola virus subtype Reston: Differences among Ebola subtypes. Archives of Virology. 2001;**146**:2021-2027

[15] Bray M, Paragas J. Experimental therapy of filovirus infections. Antiviral Research. 2002;**54**:1-17

[16] Balmith M, Soliman MES. Potential Ebola drug targets – Filling the gap: A critical step forward towards the design and discovery of potential drugs. Biologia (Bratisl). 2017;**72**:1-13

[17] Takada A, Kawaoka Y. The pathogenesis of Ebola hemorrhagic fever. Trends in Microbiology. 2001;**9**:506-511

[18] Volchkov VE, Volchkova VA, Chepurnov AA, et al. Characterization of the L gene and 5 h trailer region of Ebola virus. The Journal of General Virology. 1999;**80**:355-362

[19] Colebunders R, Borchert M. Ebola haemorrhagic fever – A review. The Journal of Infection. 2000;**40**:16-20

[20] Fauci AS. Ebola — Underscoring the global disparities in health care resources. The New England Journal of Medicine. 2014;**371**:1084-1086

[21] Calain P, Bwaka MAA, Colebunders R, et al. Ebola Hemorrhagic fever in Kikwit, Democratic Republic of the Congo: Clinical observations in 103 patients. The Journal of Infectious Diseases. 1999;**179**:S1-S7

[22] World Health Organization (WHO). Ebola Data and Statistic. 2016. Available from: http://apps.who.int/gho/data/view.ebola-sitrep.ebola-summary-latest?lang=en [Accessed: May 11, 2016]

[23] Report of a WHO/International Study team. Ebola haemorrhagic fever in Sudan, 1976. Bulletin of the World Health Organization. 1978;**56**:247-270

[24] Team WHOER. Ebola virus disease in West Africa – The first 9 months of the epidemic and forward projections. The New England Journal of Medicine. 2014;**371**:1481-1495

[25] Stein RA. What is Ebola? International Journal of Clinical Practice. 2014;**69**:49-58

[26] Briand S, Bertherat E, Cox P, et al. The international Ebola emergency. The New England Journal of Medicine. 2014;**371**:1180-1183

[27] World Health Organization (WHO). Ebola Virus Disease. 2016. Available from: http://www.who.int/mediacentre/factsheets/fs103/en/

[28] Centers for Disease Control and Prevention. 2014-2016 Ebola Outbreak in West Africa. U.S. Department of Health and Human Services. 2016. Available from: https://www.cdc.gov/vhf/ebola/outbreaks/2014-west-africa/index.html

[29] MacNeil A, Rollin PE. Ebola and Marburg Hemorrhagic fevers: Neglected tropical diseases? PLoS Neglected Tropical Diseases. 2012;**6**:e1546

[30] Trouiller P, Olliaro P, Torreele E, et al. Drug development for neglected diseases: A deficient market and a public-health policy failure. Lancet (London, England). 2002;**359**:2188-2194

[31] Qiu X, Wong G, Audet J, et al. Reversion of advanced Ebola virus disease in nonhuman primates with ZMapp. Nature. Epub ahead of print August 2014. DOI: 10.1038/nature13777

[32] La Torre G, Nicosia V, Cardi M, et al. Ebola: A review on the state of the art on prevention and treatment. Asian Pacific Journal of Tropical Biomedicine. 2014;**4**:925-927

[33] Warren TK, Wells J, Panchal RG, et al. Protection against filovirus diseases by a novel broad-spectrum nucleoside analogue BCX4430. Nature. 2014;**508**:402-405

[34] Sweiti H, Ekwunife O, Jaschinski T, et al. Repurposed therapeutic agents targeting the Ebola virus: A systematic review. Current Therapeutic Research. 2017;**84**:10-21

[35] Johansen LM, Dewald LE, Shoemaker CJ, et al. A screen of approved drugs and molecular probes identifies therapeutics with anti – Ebola virus activity. Science Translational Medicine. 2015;**7**:1-14

[36] Madrid PB, Chopra S, Manger ID, et al. A systematic screen of FDA-approved drugs for inhibitors of biological threat agents. PLoS One. 2013;**8**. Epub ahead of print. DOI: 10.1371/journal.pone.0060579

[37] Piccoli E, Nadai M, Caretta CM, et al. Amiodarone impairs trafficking through late endosomes inducing a Niemann-pick C-like phenotype. Biochemical Pharmacology. 2011;**82**:1234-1249

[38] Stanley DA, Honko AN, Asiedu C, et al. Chimpanzee adenovirus vaccine generates acute and durable protective immunity against Ebolavirus challenge. Nature Medicine. 2014;**20**:1126

[39] Henao-Restrepo AM, Camacho A, Longini IM, et al. Efficacy and effectiveness of an rVSV-vectored vaccine in preventing Ebola virus disease: final results from the Guinea ring vaccination, open-label, cluster-randomised trial (Ebola Ça Suffit!). Lancet. December 2016. Epub ahead of print. DOI: 10.1016/S0140-6736(16)32621-6

[40] Federal Select Agent Program. Select Agents & Toxins FAQ's. U.S. Department of Health and Human Services 2017. Available from: https://www.selectagents.gov/faq-general.html

[41] Federal Select Agent Program. APHIS/CDC Form 1 – Registration for Possession, Use, and Transfer of Select Agents and Toxins. U.S. Department of Health and Human Services 2017. Available from: https://www.selectagents.gov/form1.html

[42] Center for Disease Control and Prevention. Interim Guidance Regarding Compliance with Select Agent Regulations for Laboratories Handling Patient Specimens Under Investigation or Confirmed for Ebola Virus Disease (EVD)

[43] Kapetanovic IM. Computer-aided drug discovery and development (CADDD): In silico-chemico-biological approach. Chemico-Biological Interactions. 2008;**171**:165-176

[44] Scott DE, Coyne AG, Hudson SA, et al. Fragment based approaches in drug discovery and chemical biology. Biochemistry. 2012;**51**:4990-5003

[45] Halperin I, Ma B, Wolfson H, et al. Principles of docking : An overview of search algorithms and a guide to scoring functions. Proteins: Structure, Function, and Genetics. 2002;**443**:409-443

[46] Hassan Baig M, Ahmad K, Roy S, et al. Computer aided drug design: Success and limitations. Current Pharmaceutical Design. 2016;**22**:572-581

[47] Talele T, Khedkar S, Rigby A. Successful applications of computer aided drug discovery: Moving drugs from concept to the clinic. Current Topics in Medicinal Chemistry. 2010;**10**:127-141

[48] Leung DW, Borek D, Luthra P, et al. An intrinsically disordered peptide from Ebola virus VP35 controls viral RNA synthesis by modulating nucleoprotein-RNA interactions. Cell Reports. 2015;**11**:376-389

[49] Haasnoot J, De Vries W, Geutjes EJ, et al. The ebola virus VP35 protein is a suppressor of RNA silencing. PLoS Pathogens. 2007;**3**:0794-0803

[50] Brown CS, Lee MS, Leung DW, et al. In silico derived small molecules bind the filovirus VP35 protein and inhibit its polymerase cofactor activity. Journal of Molecular Biology. 2014;**426**:2045-2058

[51] Médail F, Quézel P. Biodiversity hotspots in the Mediterranean Basin: Setting global conservation priorities. Conservation Biology. 1999;**13**:1510-1513

[52] Arifin HS, Nakagoshi N. Landscape ecology and urban biodiversity in tropical Indonesian cities. Landscape and Ecological Engineering. 2011;**7**:33-43

[53] Yanuar A, Mun'im A, Lagho ABA, et al. Medicinal plants database and three dimensional structure of the chemical compounds from medicinal plants in Indonesia. International Journal of Computer Science 2011;**8**:180-183

[54] Allin SM, Duffy LJ, Page PCB, et al. Towards a total synthesis of the manadomanzamine alkaloids: The first asymmetric construction of the pentacyclic indole core. Tetrahedron Letters. 2007;**48**:4711-4714

[55] Jadulco R, Brauers G, Edrada RA, et al. New metabolites from sponge-derived fungi *Curvularia lunata* and *Cladosporium herbarum*. Journal of Natural Products. 2002;**65**:730-733

[56] Liu H, Pratasik SB, Nishikawa T, et al. Lissoclibadin 1, a novel trimeric sulfur-bridged dopamine derivative, from the tropical ascidian Lissoclinum cf. badium. Tetrahedron Letters. 2004;**45**:7015-7017

[57] Cencic R, Carrier M, Galicia-Vázquez G, et al. Antitumor activity and mechanism of action of the cyclopenta[b]benzofuran, silvestrol. PLoS One. 2009;**4**:e5223

[58] Litaudon M, Trigalo F, Martin M-T, et al. Lissoclinotoxins: Antibiotic polysulfur derivatives from the tunicate Lissoclinum perforatum. Revised structure of lissoclinotoxin A. Tetrahedron. 1994;**50**:5323-5334

[59] Litaudon M, Guyot M. Lissoclinotoxin a, an antibiotic 1,2,3-trithiane derivative from the tunicate Lissoclinum perforatum. Tetrahedron Letters. 1991;**32**:911-914

[60] Liu H, Fujiwara T, Nishikawa T, et al. Lissoclibadins 1-3, three new polysulfur alkaloids, from the ascidian Lissoclinum cf. badium. Tetrahedron. 2005;**61**:8611-8615

[61] Wang W, Takahashi O, Oda T, et al. Lissoclibadins 8-14, polysulfur dopamine-derived alkaloids from the colonial ascidian Lissoclinum cf. badium. Tetrahedron. 2009; **65**:9598-9603

[62] Nakazawa T, Xu J, Nishikawa T, et al. Lissoclibadins 4-7, polysulfur aromatic alkaloids from the Indonesian ascidian Lissoclinum cf. badium. Journal of Natural Products. 2007;**70**:439-442

[63] Oda T, Kamoshita K, Maruyama S, et al. Cytotoxicity of Lissoclibadins and Lissoclinotoxins, isolated from a Tropical Ascidian Lissoclinum cf. badium, against human solid-tumor-derived cell lines. Biological and Pharmaceutical Bulletin. 2007;**30**:385-387

[64] Pieterse CMJ, van Loon LC. Salicylic acid-independent plant defence pathways. Trends in Plant Science 1999;**4**:52-58

[65] Yan L-H, Nuhant P, Sinigaglia I, et al. Synthesis of the Indolic Pentacyclic Core of Manadomanzamine A following biogenetically based strategies. European Journal of Organic Chemistry. 2012;**2012**:1147-1157

[66] Oshimi S, Zaima K, Matsuno Y, et al. Studies on the constituents from the fruits of Phaleria macrocarpa. Journal of Natural Medicines. 2008;**62**:207-210

[67] Calderon-Montano MJ, Burgos-Moron E, Perez-Guerrero C, et al. A review on the dietary flavonoid Kaempferol Mini-Reviews Medicinal Chemistry 2011;**11**:298-344

[68] Liu T, Nair SJ, Lescarbeau A, et al. Synthetic silvestrol analogues as potent and selective protein synthesis inhibitors. Journal of Medicinal Chemistry. 2012;**55**:8859-8878

[69] Kamiya K, Tanaka Y, Endang H, et al. New Anthraquinone and Iridoid from the fruits of Morinda Citrifolia. Chemical & Pharmaceutical Bulletin. 2005;**53**:1597-1599

[70] Chemical Computing Group. Molecular Operating Environment (MOE) 2014.09. 2015. Chemical Computing Group Inc. Available from: https://www.chemcomp.com/MOE2014.htm

[71] Tambunan USF, Parikesit AA, Nasution MAF, et al. Exposing the molecular screening method of Indonesian natural products derivate as drug candidates for cervical cancer. Iranian Journal of Pharmaceutical Research. 2017;**16**:1115-1129

[72] Tambunan USF, Alkaff AH, Nasution MAF, et al. Screening of commercial cyclic peptide conjugated to HIV-1 tat peptide as inhibitor of N-terminal heptad repeat glycoprotein-2 ectodomain Ebola virus through in silico analysis. Journal of Molecular Graphics & Modelling. 2017;**74**:366-378

[73] Liebeschuetz JW, Cole JC, Korb O. Pose prediction and virtual screening performance of GOLD scoring functions in a standardized test. Journal of Computer-Aided Molecular Design. 2012;**26**:737-748